DATA AND MODELS IN ACTION

Current Issues in Production Ecology

VOLUME 5

Data and Models in Action

Methodological Issues in Production Ecology

Edited by

A. STEIN

Department of Environmental Sciences,
Agricultural University,
Wageningen, The Netherlands

and

F. W. T. PENNING DE VRIES

International Board for Soil Research and Management,
Bangkok, Thailand

SPRINGER-SCIENCE+BUSINESS MEDIA, B.V.

A C.I.P. Catalogue record for this book is available from the Library of Congress.

ISBN 978-90-481-5181-3 ISBN 978-94-015-9217-8 (eBook)
DOI 10.1007/978-94-015-9217-8

Printed on acid-free paper

Table of contents

Preface

Under leadership of CT de Wit a large amount of modeling, building prototypes and also application, was carried out in the 1970s and 1980s. Comprehensive models were built, evaluated and carefully documented in the areas of crop growth production, plant breeding, soil water and nutrients, and in crop protection. Simulation techniques and biophysical theories developed in parallel. Simulation and experimentation always went hand in hand. Much of this work is documented in a long series of PhD theses under supervision of De Wit, in the series of Simulation Monographs (PUDOC), and in numerous other publications. This work has inspired many scientists across the global science community.

The CT de Wit Graduate School of Production Ecology (PE) of the Wageningen University builds further on this platform and finds new subjects for research on and with models, and data. The PE platform provides also an excellent opportunity to develop contacts, cooperation and joint software with research groups in related fields and abroad. This book precipitates from such an exploration in new directions. We realize that modern information systems and statistics can offer a substantial contribution to the modelling framework. Good examples can be found here, and these provide a clear direction for the years to come.

This volume is based on a series of seminars in 1996 and 1997, conducted by the Graduate School of Production Ecology. We are grateful to the board and the staff of the research school for supporting this series. We thank again all speakers for their contributions, and particularly those who also converted their thoughts into the written contribution that together constitute this book. We are also very grateful for contributions by PJ Schotman and JW Schut who assisted us in organizing these seminars.

Alfred Stein
Frits Penning de Vries

Contributors & Editors

MF Goodchild
Dept. of Geography, National Center for Geographic Information and Analysis, University of California, Santa Barbara, CA 93106-4060, USA
E-mail: good@ncgi.ucsb.edu

MJW Jansen
Centre for Biometry Wageningen, P.O. Box 100, 6700 AC Wageningen, The Netherlands.
E-mail: m.j.w.jansen@cbw.dlo.nl

MH Jeuffroy
Institut National de la Recherche Agronomique, Unité d'Agronomie, 78 850 Thiverval-Grignon, France
E-mail: jeuffroy@bcgn.grignon.inra.fr

PA Longley
School of Geographical Sciences, University of Bristol, University Road Bristol BS8 1SS, United Kingdom
E-mail: Paul.Longley@bristol.ac.uk

M Molenaar
International Institute for Aerospace Survey and Eartch Sciences (ITC) PO Box 6, 7500 AA Enschede, The Netherlands
E-mail: molenaar@itc.nl

FWT Penning de Vries
International Board for Soil Research and Management (IBSRAM) PO Box 9-109, Bangkhen, Bangkok 10900, Thailand
E-mail: ibsram@cgnet.com

JU Smith
IACR-Rothamsted, Harpenden, Herts. AL5 2JQ, United Kingdom.
E-mail: jo.smith@bbsrc.ac.uk

PJ Schotman
Dept. of Computer Science and Dept. of Horticulture, Wageningen Agricultural University. Current address: Bolesian BV, Steenovenweg 19, 5708 HN Helmond, The Netherlands
E-mail: Schotman@tip.nl

A Stein
Dept. of Environmental Sciences, Wageningen Agricultural University
PO Box 37, 6700 AA Wageningen, The Netherlands
E-mail: Alfred.Stein@bodlan.beng.wau.nl

G van Straten
Wageningen Agricultural University, Systems and Control Group, Department of Agricultural Engineering and Physics, Bomenweg 4, 6703 HD Wageningen, The Netherlands.
E-mail: Gerrit.vanStraten@user.AenF.WAU.NL

JL Top
Agrotechnological Research Institute ATO-DLO, PO Box 17,
6700 AA Wageningen, The Netherlands
E-mail: j.l.top@ato.dlo.nl

P Vossen
Joint Research Centre of the European Community, 21020 - Ispra (VA), Italy
Current address: DE XXIV, Belliardstraat 232, 1049 Brussels, Belgium
E-mail: paul.vossen@dg24.cec.be

General Introduction

A Stein and FWT Penning de Vries

Production ecology like much of agricultural research depends for a large part upon the use of models. Models that describe the growth of a particular crop on a farmer's field, models that describe the movement of water through the soil, and models that describe the development of plant diseases. Much research is being done for development of models and currently even more on the use of models. Both development and use depend upon facilities for making abstractions from real-world problems. Such abstractions are nowadays rather easy to make, in particular as computer systems are readily available, and to a large extent are at the core of scientific progress.

As soon as data become available, statistical questions emerge. Important problems concern the amount of data to be collected, validation and calibration of models and sensitivity analyses. New questions, however, also include matters of scale, whereas modern statistical procedures are increasingly being applied. In this volume we will focus on the use of models in particular.

Part I sets the stage for the basic problems. The first chapter, by MH Jeuffroy, describes the development and use of a specific growth model on pea development. The model is a particular interesting one, as the description it offers is very close to actual observations: the model is therefore a very good reflection of reality. The chapter discusses the role of data for the modeling purpose and derives some general conclusions about the use of models and their careful development. A distinction is made in calibration and validation and statistical issues are discussed. The use of the model in various circumstances is presented. This model serves as a basis for the studies to follow. Key issues raised in this chapter are that experimental data are necessary to evaluate the sub-models across a range of situations where the model will be used, that the lack of fit between observed and simulated situations helps to improve the model, and that the model is a tool to quantify the effects of limiting factors, which are not included in the model. The model must be evaluated across the range of environmental conditions observed in the field.

The second chapter, by G van Straten, focuses on development of crop growth models suitable for cultivation steering. Research oriented models tend to be too large or too complex to be used directly. Various options to remedy this situation are discussed. Mechanistic models can be reduced, either by reasoning or by formal identification of methods from the original prototype model. Alternatively, mechanistic black-box models such as artificial neural nets can be derived directly from the data. In this chapter, off-the-shelf crop growth models are applied instead, from which a reduced big-leaf-big-fruit tomato model is derived. The chapter provides some general procedures for tuning specific models towards various circumstances. It extends chapter 1 by adapting models by various statistical and mathematical themes. An important issue in this chapter concerns handling of mismatch, ranging from modeling the error to various procedures to reducing the error. Of primary concern is the relation between deterministic and stochastic modeling. Key statements in this chapter are that more often than not the validity domain of a model is not well documented. Enhancing deterministic models with a stochastic part does not reduce uncertainty, and it reduces the bias in parameter estimates and increases the reliability of the forecasts and that, if detailing of models is unavoidable, cybernetic modeling may be an interesting option to solve the problem of resource allocation.

The third chapter, by P Vossen, addresses yield forecasting at the European scale. It addresses the important issue of upscaling. We will come across this theme as well in chapters 5 and 7, but then from a different perspective. Crop models, developed for single field conditions are applied at the European scale. Various new problems are encountered, which in particular relate to i) the scale at which the models are being developed, usually the field scale, and ii) the scale of application, here 50 × 50 km blocks. In particular, attention focuses on estimation for global radiation. It is discussed how standard models and empirical relations can be combined to give the best results. Statistical modeling can overcome part of the problems, but puts a large emphasis on data availability. A balance between stochastic and deterministic models appears to be the most suitable one. The chapter concludes with a list of crop databases available for general purposes. Key issues in this chapter are that information is often available at the wrong scale, that the time trend component is often more significant than the agrometeorological model outputs, and that introduction of remotely sensed information into models will significantly improve spatial representativity and trustworthiness of both model inputs and outputs.

Part II addresses statistical questions in validation of models and change of scale. Chapter 4, by MJW Jansen, presents a Bayesian method for validation of models. Model validation has to do with finding the best possible set of values for a range of circumstances. This in turn defines the domain of applicability for a model. The model must be applicable within specified domain limits, whereas application outside the domain may lead to highly erratic outcomes. Domains should be provided together with the model. Many procedures exist to calibrate models. Bayesian procedures have gained a wider popularity, as they allow incorporating subjective and personal experience, and provide an excellent starting point for uncertainty analysis and decision support. Bayesian calibration involves either a grid method or Monte Carlo methods. In an example on the SUCROS model, the use of measurements from a previous year yields a substantial improvement.

Chapter 5, by JU Smith, focuses on up- and downscaling. Up- and downscaling has a rather broad range of meanings nowadays. Upscaling in space applies a model developed for a small area of land to a large area of land, whereas downscaling applies a model developed for a large area to a small piece of land. The meanings of the words 'small' and 'large' are context dependent. Upscaling in time applies a model developed for a small time interval to a much larger interval, and downscaling applies a model for a large time interval to a small interval. Upscaling in complexity derives a general model from several more complex and more detailed models, downscaling means that general models address small, often disciplinary problems. In this chapter general procedures are described, that are applicable as well for validation purposes. Scale is treated in this chapter mainly as the arithmetic definition: as the spatial resolution, for which the model is *derived*, compared to the spatial resolution for which the model is *applied*. Several measures for model evaluation are defined. The SUNDIAL model, a dynamic nitrogen turnover model, is analyzed as an example. The scale of the model should be driven by data availability and the questions asked. Two practical reasons for changing scale emerge: transferability of the model to novel environments and model robustness. The scope of the model is refined by omitting any processes that the sensitivity analysis suggests are redundant at the larger scale. Ensuring that an independent module characterizes each process requires individual evaluation of process modules. The paper presents the following statements:

- discontinuously heterogeneous variables are most conveniently scaled-up by describing them as the majority value;
- a simple minimum information requirement (MIR) model is useful for application at the catchment scale;
- weather data may be interpolated from regional meteorological offices by using spatial regression downscaling together with a local stochastic weather generator.

Part III discusses the use of Geographical Information Systems (GIS). Chapter 6, by PA Longley and MF Goodchild, presents linking between models and GIS. In this chapter they describe how and why model linkage requires understanding of the digital data that provide the infrastructure to analysis. The computer has evolved from an equation-solving and data-handling device to a beginning-to-end environment for the modeling of complex systems and processes. This becomes in particular relevant in modern scientific research. GIS provide us with means of tailoring multi-purpose generic data to the context of particular applications. They state that the rise of interdisciplinary science is leading to the situation in which the creators of spatial data may be widely separated from some of the communities of end users. On the other hand, requirements are being imposed on the quality of the data. The question is raised whether the new digital data infrastructure can be assembled together in a sufficiently accurate, orderly and rational way to bridge relevance, richness and academic respectability. They conclude with stating that it is a major challenge to the GIS community to devise methods to reconcile diverse data sets with different data structures or spatial referencing systems. Only in this way will GIS be able to tease out the complex relationships that exist between projects, data sets, and analytic techniques in modern science.

In chapter 7, M Molenaar addresses multi-scale approaches for geodata. Geodata are data with a spatial reference, and hence both modeling for areas of land and issues of upscaling are of particular interest here. Multi-scale approaches are at present a focal point of GIS research because awareness is rising that many processes on the earth surface can only be monitored and managed if they are understood in their geographical context. Several approaches to generalizations are addressed and several examples illustrate the use of multi-scale approaches for modeling for production ecology. In this chapter, a distinction is made between class generalization and arithmetic forms of upscaling, with an emphasis on generalization procedures. Four different strategies are distinguished: geometry driven generalization, class driven generalization, functional generalization and structural generalization. Elements of graph theory are included, as are elements of database structures. The chapter is illustrated with two examples, one with a hydrological content and one on erosion.

Part IV addresses some interesting developments for the future. Chapter 8 by PJ Schotman is on object oriented programming. This chapter discusses some examples of object-oriented design in agricultural research. It describes how at a somewhat more abstract distance object-oriented programming can serve as a tool for developing agricultural models. Attention focuses on short procedures and statements, in stead of on long programs or programming segments, hence allowing a greater flexibility and enhanced communication facilities. Object-oriented programming languages have extended functionality as compared to their conventional languages. The main advantage will be its reuse by people other than the developers. As an example, the chapter deals with the 'composite' design pattern as a basis for scaling.

Chapter 9, by J Top, is on automated modeling. The author expects that automated modeling will exploit information technology to turn the art of modeling into a profession. Explicit bookkeeping of the modeling assumptions will be a major step forward in the availability of high-quality models. Some trends in automated modeling are identified. Attention focuses primarily on trends in hardware and software. Benefits and possibilities are discussed.

I Practical Aspects of Modeling

Part I contains general practical aspects of modeling for production ecology. A central role is given to modeling of individual plants, but elements of data availability, scale, model extension and reliability already show up. It is divided into three chapters. The first chapter gives an impression of a particular, good working, way of modeling peas, and serves as an illustration throughout the book. The second chapter describes procedures for crop modeling in cultivation control, and hence already addresses specific aims of modeling. The third chapter discusses elements of upscaling. In particular attention is given to the important aspect of meteorological data. This part gives as a background an overview of those processes that will play a substantial role throughout the remainder of the book. Fundamental principles of methodology for production ecology are presented and illustrated throughout.

1 Development of a morphogenetic model from field and lab data: modeling the seed number per node on a pea stem

MH Jeuffroy

1.1 Introduction

Crop models can be used for many purposes (Boote et al 1996). They may help to interpret experimental results, to understand interactions of numerous factors in crop growth and production, and to support farmers in decision making. For these different aims, models must give a realistic representation of the complex reality. On the one hand, a crop growth model must be comparable to data from the field, e.g. by giving a good account of the reality. In addition, it must allow to take results from experimental data into account, so that description of the natural phenomena (model structure and parameters) can be derived from observation rather than from speculation (Monteith 1996).

In this chapter, several relations between a model (its building and its use) and data will be illustrated. The model simulates the seed number on each reproductive node of a pea stem (Jeuffroy 1994).

A pea plant has a main stem that can bear branches, the number of, which can be highly variable. On each stem, from the base to the top, one first observes some vegetative nodes with only a composed leaf. Then, from the first flowering node, which is about the 15th node on the cultivar Solara, each following node bears a composed leaf, and one or several flowers, which will give pods containing seeds, if they do not abort. Flowers appear successively on different nodes and development and growth of pods and seeds is sequential. Hence various organs at different ages can co-exist:

- vegetative organs in development (as the apex continues to develop during the reproductive period);
- young flowers (on the upper nodes)
- young pods below and
- old pods with filling seeds on the first reproductive nodes.

Many reproductive organs (flowers, pods and seeds) generally abort on a pea stem. The proportion of aborted organs is not constant among the different nodes, resulting in a heterogeneous distribution of seed numbers among the different nodes, the so-called profile of seed number. As development is sequential, the seed number on each node is not fixed at the same time. Thus the seed formation on the successive nodes is not realized in the same environmental conditions. For example, a limiting factor occurring early will reduce the seed number on the first nodes; if it occurs later, it will only have an effect on the last nodes of the stem. Moreover, it can be assumed that seed formation on the last nodes depends on what happened on the first nodes. Thus in

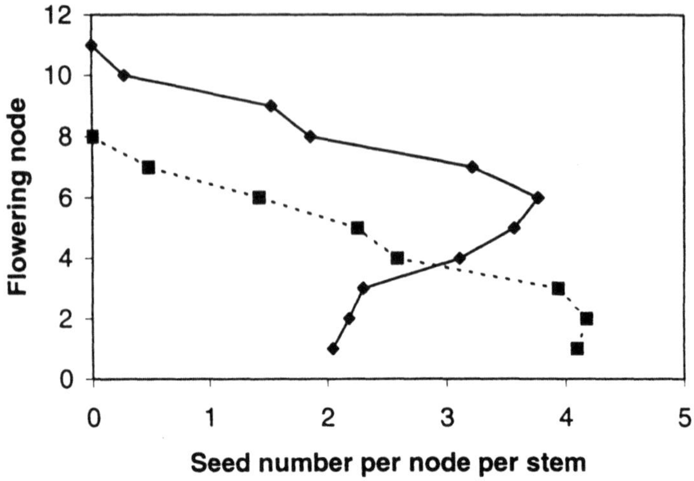

Figure 1.1. The two main forms of profiles of seed numbers per node observed in the fields.

many situations, the total seed number per stem can be understood only if we understand the seed formation on each node.
In farmers' fields, various profiles are observed. They have mainly two typical forms (Fig. 1.1). Some are rather high, with a high number of flowering nodes, and generally bear few seeds on the first reproductive nodes; others are low, with few flowering nodes, and a high number of seeds on the first reproductive nodes. In both cases, the seed number decreases regularly on the upper nodes. The total seed number per stem is linked to the form of the profile, and the highest seed numbers per stem generally occur on rather low profiles. To understand the high observed yield variability within farmer's fields, mainly linked to variability of seed numbers (Doré 1992); it is thus essential to understand the diversity of these profiles.

In the literature, the main hypothesis to explain the profiles is nutritional: as there are many sinks on a stem, of various kinds and ages, there is a high competition for assimilates between them, and particularly between pods, explaining their different seed numbers. Because the organs involved in this highly complex competition are numerous and from various nature, it may be useful to make a model including assimilate partitioning. We thus assume that the variability of the profiles of seed numbers per node observed in the fields can be understood with a pea growth model.

The model is a tool to understand the diversity observed in the fields.

1.2 The structure of the model

1.2.1 Seed and plant development

In the pod life, three main developmental stages can be defined: *flowering*, *final stage in seed abortion* and *physiological maturity*. Between *flowering* and *final stage in seed abortion*, cell divisions occur in the ovules (Ney et al 1993). During this period, a whole pod or some of the ovules contained in it may abort, reducing the final seed number of the pod. After the *final stage in seed abortion* the seeds cannot abort any more in the pod (Pigeaire et al 1986) and the seed number is fixed. This stage corresponds to the beginning of seed filling (Ney et al 1993).

If we focus upon the stem scale we notice that progression of these three main stages along the stem is linear in cumulative degree-days (Ney and Turc 1993). Duration between occurrence of a stage on two consecutive nodes is constant, and equals ± 45 degree-days for flowering and ± 40 degree-days for final stage in seed abortion in cv. Solara. It is thus possible to know the age of each pod of the stem, at each day of the reproductive period, as soon as the date of beginning of flowering on the stem is recorded. Moreover, the period of seed number formation can be defined between flowering of the first flower (FLO1) and final stage in seed abortion of the last reproductive node (FSSA). This will be the period considered for the model.

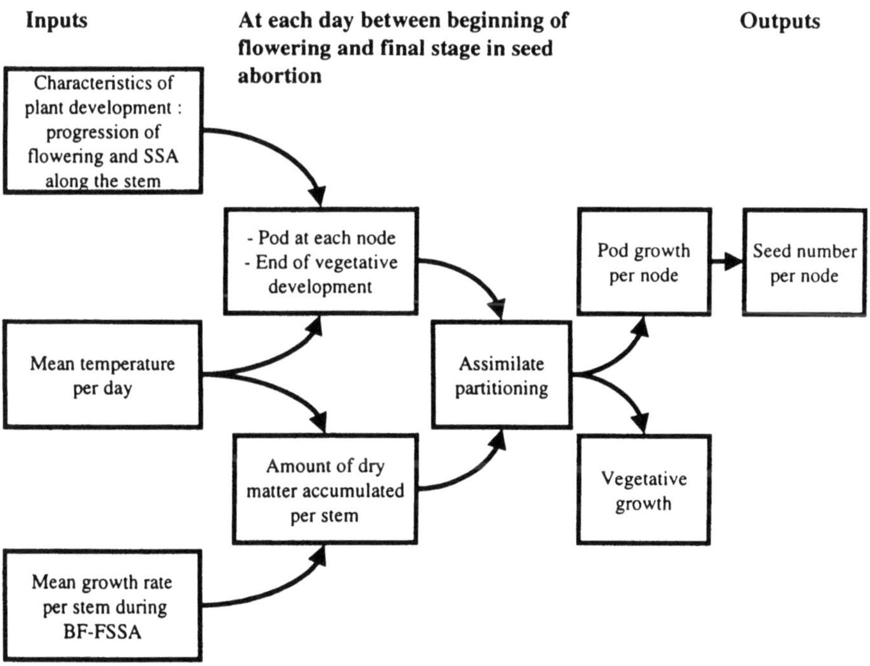

Figure 1.2. Structure of the model (according to hypothesis of plant functioning). BF = beginning of flowering : FSSA = final stage in seed abortion.

1.2.2 The model framework

The original modeling assumption is that the seed number of a pod depends on the pod growth dynamics before it reaches its final stage in seed abortion (Fig. 1.2). This was demonstrated for individual pods from all nodes, on various situations in the field (Jeuffroy and Chabanet 1994). Then growth dynamics of each pod depends on the whole stem growth dynamics and on the pattern of assimilate partitioning among all sinks of the stem along time. Assimilate partitioning may then depend upon the nature and age of the various sinks in competition, which are mainly developmental variables.
The inputs of the model are thus

- characteristics of plant development
- mean daily temperature to simulate pod age at each node and the end of the vegetative development, and
- mean growth rate per stem to estimate the amount of available assimilate at each daily step.

The model is thus composed of three main modules

- estimation of seed number per pod according to early pod growth
- assimilate partitioning between sinks, and
- determination of the end of flowering (fixing the final number of flowering nodes).

I will focus on the module of assimilate partitioning.

1.3 The module of assimilate partitioning

1.3.1 The hypothesis of assimilate partitioning

Assimilate partitioning is a weak point of many crop models (Whisler et al 1986). Several analytical results exist to build such a module, being often contradictory or insufficient. Also, some questions arise from literature. The first concerns the possibility of internodal translocation of assimilates during the period of seed formation. Some physiological studies showed that most of the assimilates allocated to a pod came from the subtending leaf (Flinn and Pate 1970) while other studies showed the possibility for a pod to receive assimilates even if its subtending leaf did not produce assimilates (Szynkier 1974). If it is possible to transfer assimilates between nodes, the next question concerns the existence of a hierarchy in the carbon distribution among the various sinks, some of them having priority on others. Finally the rules of assimilate partitioning among sinks during the whole period of seed set must be determined. To answer these questions, which are the structural assumptions of the model, several experiments were realized.

Data are necessary to test the structural assumprions of the model.

1.3.2 An experiment to test these hypotheses

A first experiment analyzed the possibility of assimilate transfer among nodes, and of existence of general determinants for assimilate partitioning. It was thus necessary to work on short periods and not only with biomass balance sheets, because of the high variability between plants. The experiment was carried out in a controlled environment, with ^{14}C-labeling. For the question of assimilate translocation, some individual nodes were shaded, the whole plant was fed $^{14}CO_2$, and the pods from the shaded nodes were analyzed (see Jeuffroy and Warembourg 1991, for details). For the question of priority of some sinks, control plants and partially shaded plants were given $^{14}CO_2$ (Jeuffroy and Warembourg 1991), the latter producing only one-fourth assimilates recovered from controls.

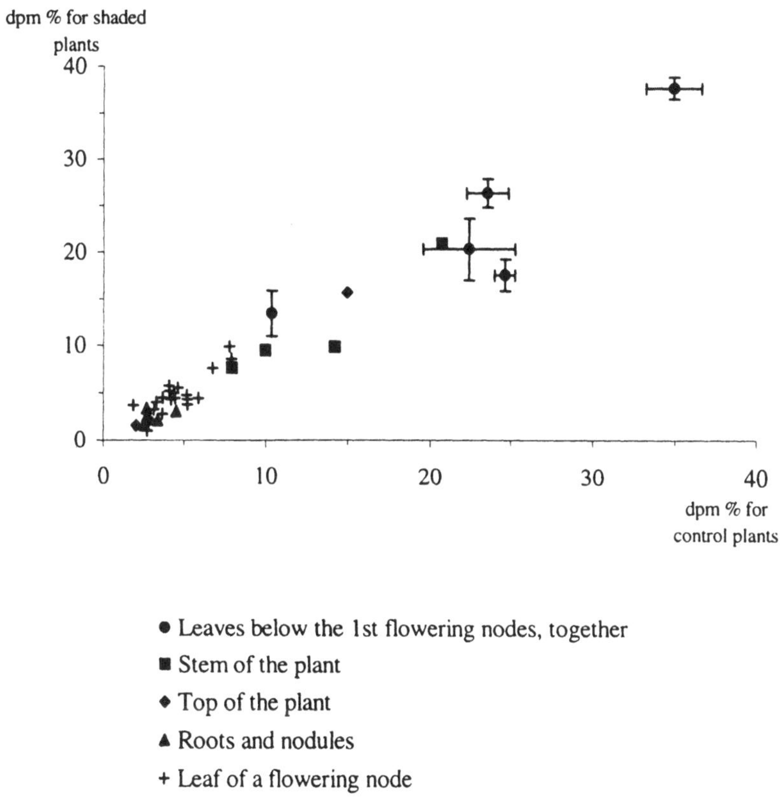

Figure 1.3. Relative distribution of ^{14}C-assimilates among vegetative organs (dpm % of total recovered in plant). Comparison between control and shaded plants. Bars indicate standard errors (only for the high % dpm). (From Jeuffroy and Warembourg 1991).

1.3.3 Some results

Assimilate translocation between nodes was demonstrated (Jeuffroy and Warembourg 1991), for several nodes and at various stages. As concerns hierarchy, we noted that the amount of assimilates recovered in the different vegetative organs from the shaded plants was very low compared to the control. The proportion of ^{14}C, however, recovered in one vegetative organ as compared to the total amount in the plant was the same between shaded and controls plants (Fig. 1.3). Thus the proportion of assimilates allocated to one vegetative sink does not vary according to the total amount of available assimilates in the plant.

Less obvious results were derived for reproductive organs, caused by high variability between pods (Fig. 1.4). It was not possible to show an effect of the total amount of produced assimilates on the proportion allocated to each pod. Taking a more detailed look at the pods, we could link the proportion of ^{14}C allocated to one pod either to its dry matter relative to all pods or to its relative seed number, according to the stage of the pod, i.e. before or after its final stage in seed abortion. Thus, in the

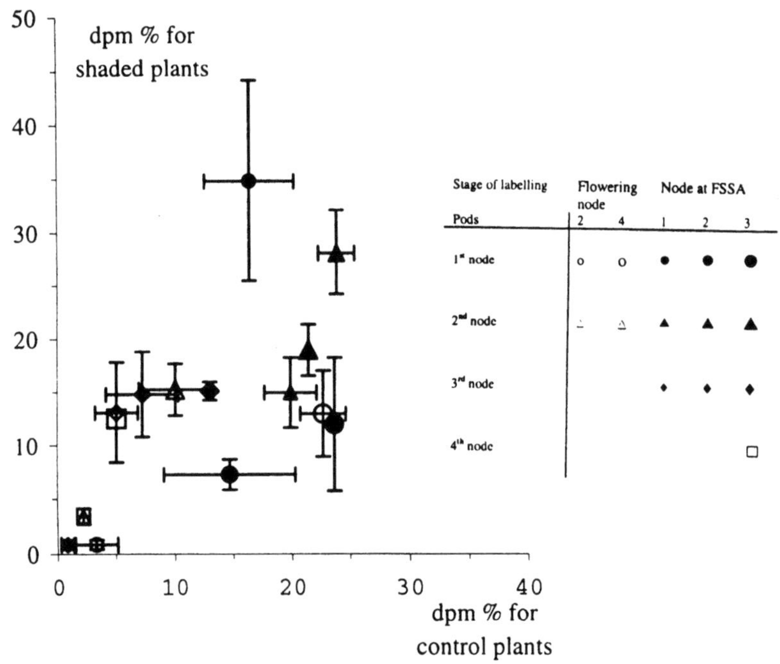

Figure 1.4. Relative distribution of ^{14}C-assimilates among pods (dpm % of total recovered in plant). Comparison between control and shaded plants. Bars indicate standard errors. (From Jeuffroy and Warembourg 1991).

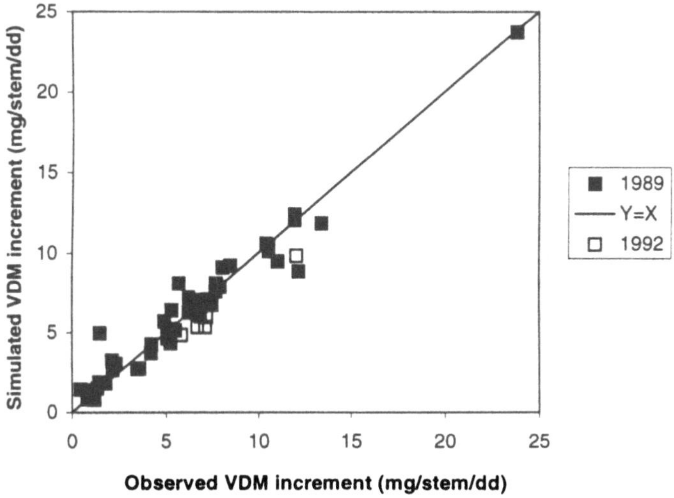

Figure 1.5. Observed and simulated vegetative dry matter increments (VDM), calculated between two successive sampling dates during the period of seed set, on the situations from Table 1.1.

competition between vegetative and reproductive sinks we could not determine priority of some sinks over others. The amount of assimilates allocated to each sink depends on its demand, proportional to a) the pod dry matter (for pods before final stage in seed abortion), b) the pod seed number (after this stage), and c) the difference between the final number of flowering nodes and the number already developed (for vegetative organs, Jeuffroy 1991).

The calculations of the amounts of assimilate allocated to each sink at each step then required the estimation of the parameters for these demands.

The experiment is a source for the estimation of the model parameters

1.3.4 Estimation of the model parameters

Estimation of model parameters required another experiment, as it was better to estimate these parameters directly in the field, where the model had to be used. In this experiment, we maintained some particular conditions to measure the potential growth of a pod. To prevent competition on the stem we cut off the apex of field grown plants

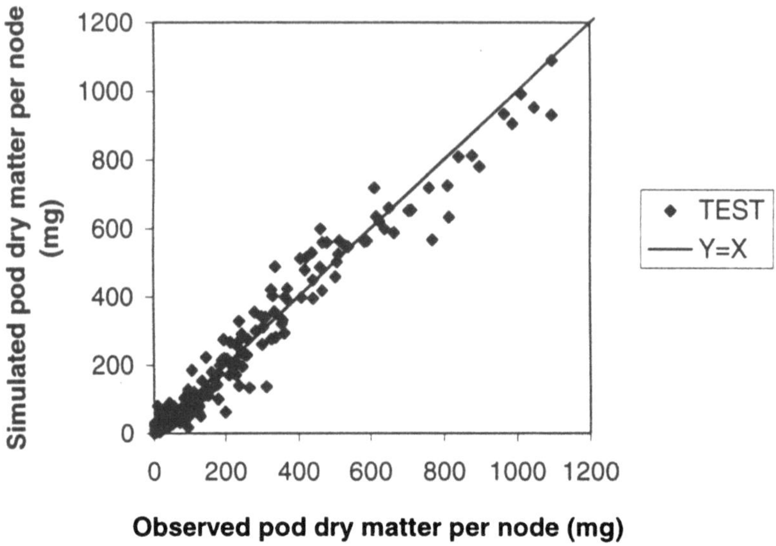

Figure 1.6. Observed and simulated pod dry matter per node, at each sampling date during the period of seed set, on the situations from Table 1.1.

after development of 4 flowering nodes and left a single pod on each stem. We measured individual pod dry matter, which was adjusted to an exponential model in the first part of the life (before final stage in seed abortion) and to a linear model in the second part (between this stage and physiological maturity). This curve gave us the parameters for pod demand (Jeuffroy and Devienne 1995). The same principles were applied for the estimation of the vegetative demand parameters (Jeuffroy 1991).

1.3.5 Evaluation of the assimilate partitioning module in the field

With these two experiments (labeling and parameter estimation), the rules for assimilate partitioning were determined. It was necessary to verify these under conditions where the model had to be used, because the plants obtained in controlled environments in the first experiments were very different from those observed in the field. Thus, another experiment was realized in the field, different from that used for parameter estimation (Jeuffroy and Devienne 1995).

Experimental data are necessary to evaluate the sub-models across a range of situations where the model must be used.

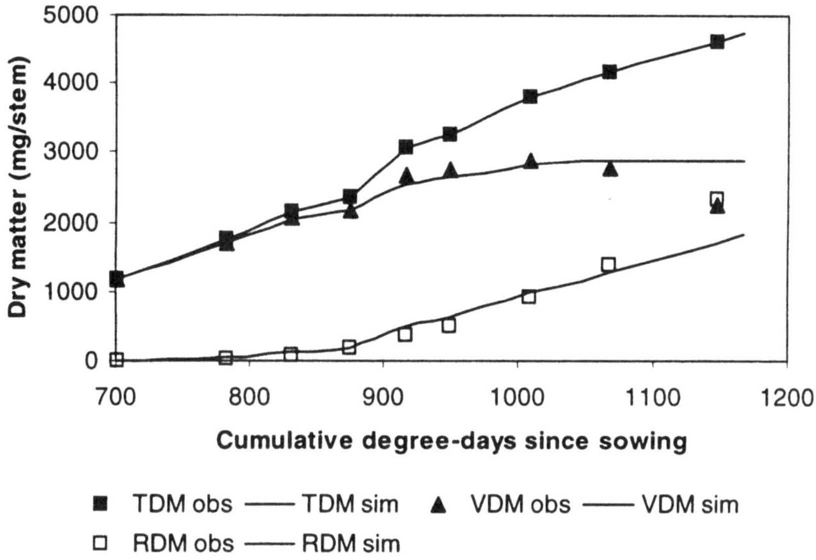

Figure 1.7. Observed and simulated growth dynamics of vegetative and reproductive parts of a peas plant. VDM = Vegetative Dry Matter, RDM = Reproductive Dry Matter, TDM = Total Dry Matter; obs = observed; sim = simulated.

Evaluating this module of assimilate partitioning in a large range of situations to be covered by the whole model, we got various sowing dates and densities during two years, inducing a large range of inputs of the model (Table 1.1). The mean growth rate per stem ranged from 5 to 14 mg·stem^{-1}·degree-day^{-1}, and the final number of flowering nodes from 4 to 10. Variability existed also in the developmental parameters. There was a good fit between observed and simulated vegetative dry matter increments, calculated between two successive sampling dates (Fig. 1.5), and also for pod dry matter at each sampling date (Fig. 1.6). The model gave also a good simulation of the growth dynamics of vegetative and reproductive organs (Fig. 1.7).

Table 1.1. Characteristics of the different crops used to test the model. (From Jeuffroy and Devienne 1995).

	Sowing date	Nr. of stems m^{-2}	Mean growth rate ($mg \cdot stem^{-1} \cdot day^{-1}$)	S_F	$1/R_F$	r^2 *a*	S_S	$1/R_S$	r^2 *b*	Final nr. of flowering nodes per stem
1989	14-3	117	9.47	756	68.5	.99	1019	40.0	1.00	6
		156	8.19	727	79.5	.98	1022	35.0	*c*	5
		257	5.16	785	54.0	.99	1022	54.0	*c*	4
	31-3	61	14.30	775	36.5	.98	943	41.0	.96	7
		107	8.87	773	35.5	.99	935	35.0	.94	6
		194	5.89	778	49.0	.99	948	39.0	*c*	5
	19-4	81	12.63	729	36.0	.91	908	68.0	.92	6
		141	7.80	705	46.0	.95	908	59.5	.90	5
		175	5.77	711	40.0	.95	910	35.0	*c*	5
1992	24 -2	67	12.98	860	37.5	.98	1068[d]	45.0[e]		10
		97	10.60	852	35.9	.98	1060[d]	45.0[e]		9
		159	7.56	841	42.3	.97	1049[d]	45.0[e]		9

S_F: date of flowering of the first node (in cumulative degree-days from sowing)
R_F: rate of progression of flowering (in node degree day^{-1})
S_S: date of final stage in seed abortion on the first node (in cumulative degree-days from sowing)
a: coefficient of correlation of the adjustment of the number of nodes, which were flowered with cumulative degree-days since sowing
b: coefficient of correlation of the adjustment of the number of nodes, which had passed FSSA with cumulative degree-days since sowing
c: only 2 observations were available
d: estimated from mean S_S-S_F of 1989
e: estimated from $1/R_S$ of 1989

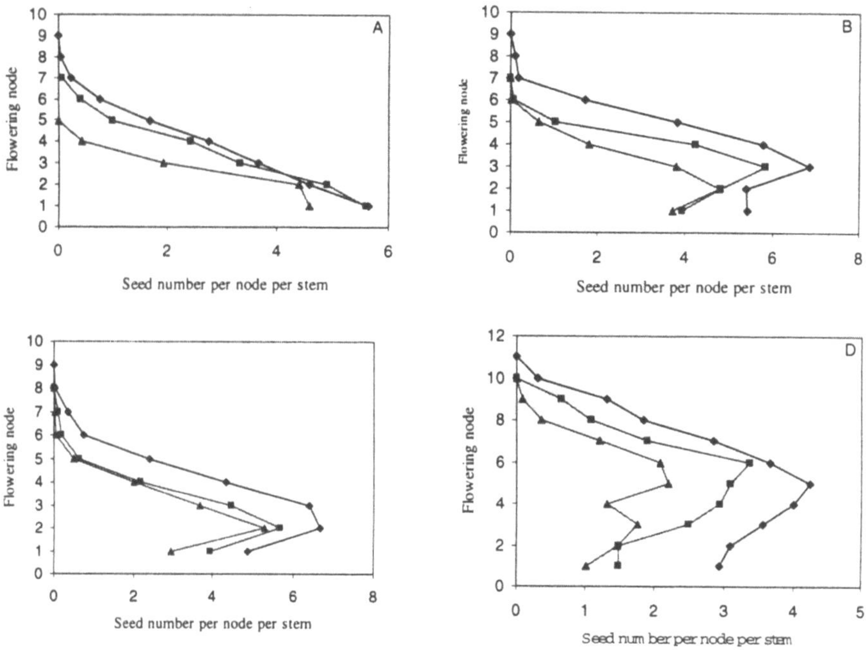

Figure 1.8. Observed profiles of seed numbers per node on the situations from Table 1.1. (From Jeuffroy and Devienne 1995).
A: sowing 14 March 1989, 117 stems m^{-2} (▲), 156 stems m^{-2} (~), 257 stems m^{-2}(■)
B: sowing 31 March 1989, 61 stems m^{-2}(▲), 107 stems m^{-2}(~), 194 stems m^{-2}(■)
C: sowing 19 April 1989, 81 stems m^{-2} (▲), 141 stems m^{-2}(~), 175 stems m^{-2}(■)
D: sowing 24 February 1992, 67 stems m^{-2}(▲), 97 stems m^{-2}(~), 159 stems m^{-2}(■)

1.4 Analysis of the various profiles of seed numbers per node observed in the fields with the model

The initial aim for model building was to understand the diversity of the profiles of seed numbers per node observed in the field. To test the ability of the model for giving a good account of this range, we evaluated the model for a large range of field situations. This step also aimed to determine the range of validity for the model.

The model must be evaluated across the range of environmental conditions observed in the fields

The model was evaluated on the range of sowing dates and densities presented in Table 1.1, resulting in a large range of in- and outputs. Fig. 1.8 shows the observed

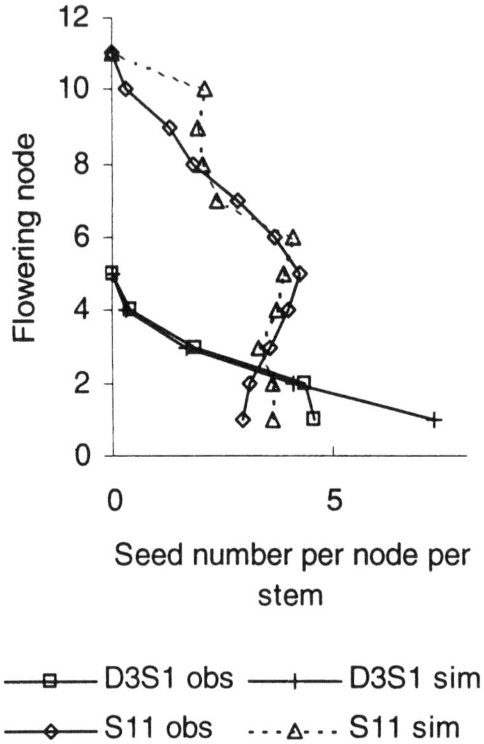

Figure 1.9. Comparison of the simulated and observed profiles of seed numbers per node on two situations from Table 1 (S11 and D3S1). S11 = sowing 24 February 1992, 67 stems m^{-2}, D3S1 = sowing 14 March 1989, 257 stems m^{-2}, obs = observed, sim = simulated

profiles of seed number for the tested situations: some are high with few seeds at the bottom, while others are lower with many seeds at the base.

The analysis can then be further detailed in considering different numbers of flowering nodes and mean growth rates, cases 3 (D3S1) and 10 (S11) of Table 1.1, respectively. There is a good fit between the observed and the simulated profiles in the two cases (Fig. 1.9).

The two different forms of profiles can be interpreted with the model. The higher number of reproductive nodes on S11 are caused by a better nitrogen nutrition status of the crop at the beginning of flowering (Jeuffroy and Sebillotte 1997). Therefore the vegetative development and growth continued during a longer period on this treatment. In the competition between vegetative organs and the first growing pods, a small amount of assimilates was allocated to the pods, in comparison to their demand. This resulted in a small number of seeds on these nodes, when compared to

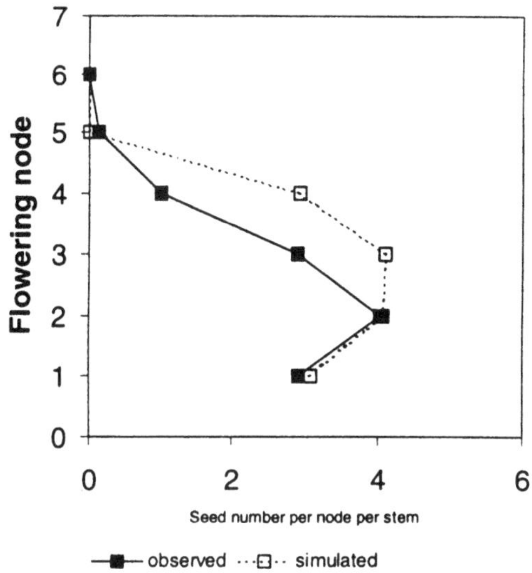

Figure 1.10. Comparison of the simulated and observed profiles of seed numbers per node on one situation in 1991.

the same nodes on D3S1, for which the competition between vegetative and reproductive organs was much shorter. Later in the cycle, competition for assimilates among pods resulted in a higher amount of assimilates allocated to the older pods, reducing the seed number on the upper last ones.

As the model well explained the observed profiles for the other tested situations, it can be concluded that it helps to understand the effects of a variety of tested inputs. Yet some are situations still not explained by the model. For example, the model generally overestimated the seed number on the last nodes, probably because it is necessary to consider the filling seeds as having a priority in assimilate distribution, as it was shown on soybean (Munier-Jolain 1994). The lack of fit then helps to improve the model. For this model, some modules are still worked upon, for example estimation of the vegetative demand, and prediction of the final number of reproductive nodes. A new model has been developed since and evaluated (Roche et al 1998).

Lack of fit between observed and simulated situations helps to improve the model

1.5 Analyzing the effect of limiting factors with the model

The model already allows us to proceed, and to estimate the effect of limiting factors, which are not included in the model. For example the simulated profile of the model during the year 1991 was different from the observed one, particularly on the nodes 3

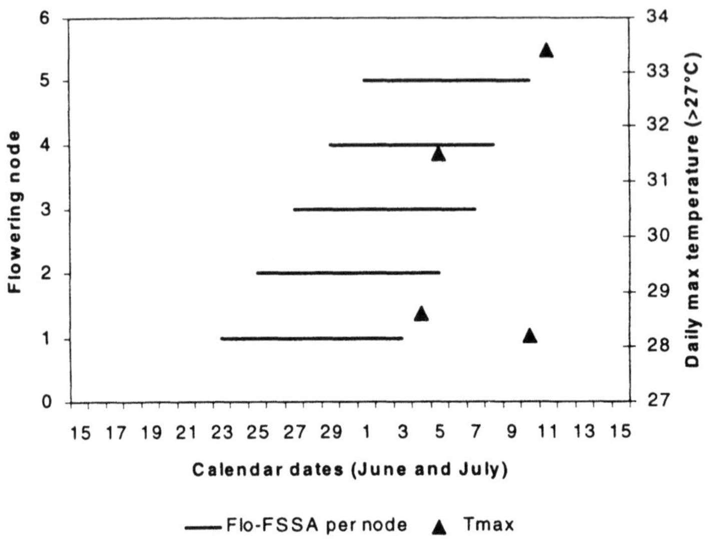

Figure 1.11. Comparison of the periods of seed formation on each node and the days with high maximum temperature.

and 4 (Fig. 1.10). Several days with very high maximum temperature occurred during the period of seed set and it is known from the literature that high temperatures during seed set cause seed abortion (Lambert and Linck 1958; Karr et al 1959) and that the period of highest sensitivity for the pods is the second half of the period between flowering and FSSA (Jeuffroy et al 1990). Upon comparing the days with high temperature to the period of seed formation on each node, it appeared that these days occurred during the period of sensitivity of the nodes 3 and 4 (Fig. 1.11), and hence could explain the gap between the observed and the simulated profiles. The model then enables to quantify the yield loss due to the limiting factor provided that good quality observations are available.

The model is a tool to quantify the effects of limiting factors, which are not included in the model.

1.6. Conclusion

In this description of building and use of a model for seed number per node on pea, several relations between data and model were illustrated.

1. The model is a tool to understand the diversity observed in the fields.

2. Data are necessary to test the hypothesis of plant functioning, giving the model structure.
3. The experiment is a source for the estimation of the model parameters.
4. Experimental data are necessary to evaluate the sub-models in the range of situations where the model must be used.
5. The lack of fit between observed and simulated situations helps to improve the model.
6. The model gives an account of the effects of limiting factors, which are not included in the model, and quantifies their effects.

Exercises

1. Consider figure 1.1. Give an interpretation of the observed differences. To describe variation in the field, one may have to model these curves. Common research questions are:
- which fraction of one model is observed,
- do two varieties sufficiently describe the variation in the field
- are there really two species, or do we observe differences for one particular crop.

2. Figure 1.3 gives an impression of the relation between dpm for control plants and for shaded plants.
- Give an interpretation of the error bars
- What could explain this range in errors? And how could it be reduced?
- A clear distinction can be observed between roots and, for example, leaves below the 1st flowering node. Explain this difference.
- Is there any indication that shaded plants provide less biomass than control plants?
- If we compare fig. 1.3 with fig. 1.4: would that give a different conclusion?

3. Figure 1.5 gives the relation between two successive sampling dates. Give a mathematical relation between these two.

4. Table 1.1 gives the characteristics of different crops used to test the model. What can we conclude with respect to the 7 different variables? Consider carefully what happens among plants with the same sowing data, and between plants with different sowing dates (take averages). Why are 1992 and 1989 difficult to compare?

5. Give your comments to each of the conclusions listed in section 1.6. Explain why you agree or disagree with the statements. Explain as well how potential problems can be overcome. For example: what has to be done in if no experimental data are available to evaluate sub-models (conclusion 4).

2 Pathways in crop modeling for cultivation control

G van Straten

2.1 Introduction

Crop growth models have been developed to gain insight in the dynamic behavior of crops, and to make predictions about potential crop yield under various circumstances. In this chapter we consider the use of crop growth models for cultivation steering. Cultivation steering is of interest in protected environments, i.e. in greenhouses, to balance crop yield versus costs of inputs - like heating costs - but also for quality control and to control the timeliness of crops on the market. Control of cultivation is also of interest in open field crops, e.g. to balance crop yield versus inputs of fertilizers, crop protection measures and environmental damage in precision farming.

Cultivation steering requires that the behavior of the crop can be predicted in the future. Models are the most versatile tool to this end. The intended use for control, however, sets a number of requirements. An important condition is that a clear distinction is made between input variables and state or output variables. Also, the model should not contain or require inputs that cannot be measured with reasonable effort (e.g. sky temperature). The model must be self-consistent, i.e. it should not require inputs that depend upon the state itself. For example, Leaf Area Index LAI is unacceptable as input, except perhaps in a feed-back situation, as it depends on the state. The model outputs must be expressed in terms that are important for the final goal. Also, it must be valid over the intended range of control. And for optimization purposes and on-line control purposes, the model must not be unduly complex.

Efforts in crop modeling in the past decades have lead to useful mechanistic, deterministic models, but they usually do not fulfil the conditions desirable for control. The aim of this chapter is to discuss possible pathways to remedy this. Two major issues are addressed:

1. how to get reduced models: mechanistic and non-mechanistic;
2. how to handle model mismatch without further reductionistic detailing of the model: the introduction of stochasticity, and mechanisms to cope with time variability and adaptation.

2.2 Off-the-shelf crop growth models

Many crop-growth models are mechanistic and deterministic. They have been developed for research purposes and are used for prediction of potential crop yield. The pathway followed to arrive at these models can be characterized as the 'reductionistic' approach. Separate experiments are set up to isolate parts of the crop growth and development, and to build sub-models for these. Next, the various sub-models are combined. Further details are added if the overall performance is not satisfactory, either by detailing sub-processes, or by adding other sub-processes.

Typical examples of such models are SUCROS (Spitters et al 1993) and LINTUL (Spitters 1990) for field crops, and TOMGRO (Dayan et al 1993), HORTISIM and others (De Koning 1994) for fruit crops like cucumber, sweet pepper and tomato in greenhouses. SUCROS and LINTUL are models with a limited number of state variables (10 and 5, respectively). Their aim is to describe different development stages of the plant. On the other hand, they are fairly complex, being in fact hybrid models, where the structure changes discontinuously as a function of the development stage state variable (see Fig. 2.1). The greenhouse models are complex models containing many state variables, in an attempt to describe the growth and development of trusses or even individual fruits. All these models are set up as discrete time models, with a time step of one day, so that day-by-day variation is averaged out.

The direct applicability of such models for cultivation steering is limited. In particular, model complexity makes them hard to use in on line control or in optimization calculations. We will investigate possible remedies in the next section. In addition, application of generalist models to a particular situation is problematic, because actual conditions usually differ from the assumptions underlying the general parameterization. This will be considered in sections 4 and further.

2.3. Model reduction and meta-modeling

To describe model reduction and meta-modeling, we will consider a complex model and assume that its predicted behavior is actually an accurate image of reality in a given validity range. We can then try to find reduced models, with a similar behavior, but more simple and hence easier to handle. There are two basically different procedures to

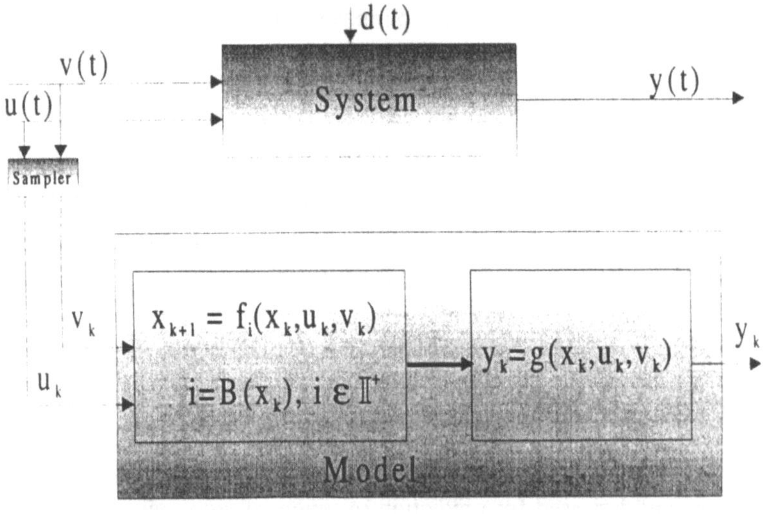

Figure 2.1. Crop (hybrid) system model. Inputs are *u*, *v*, and *d*: controlled, measured environment and disturbance inputs, respectively. States are *x*, outputs *y*. A subscript *k* indicates discrete time t_k, *i* is a positive integer, selected by the switching function *B*.

achieve such reduced models.

The first is to reduce the equations by lumping, aggregating, simplification and pruning of the original models by reasoning. As an example, Tap and Van Straten (1995) developed a big-fruit-big-leaf approximation of the tomato model with more than 300 states developed by De Koning (1994). This was done mainly by lumping individual fruits and leaves, while maintaining the basic structure. The final model structure is given in Fig. 2.2.

In addition, the model was made continuous, rather than discrete on a daily basis, for use in an optimal controller setting. This was done by introducing a dynamic assimilate buffer, and formulating a smooth switching function to prevent fruits and leafs from drawing upon assimilates when the buffer is empty. The model, therefore, still accommodates sink limited and source limited growth. Sink limited growth occurs when the assimilate demand is less than the photosynthetic production. Source limited growth occurs when the assimilate production is less than the demand from leafs and fruits. Due to the continuization, however, the system can be source limited during the night, and sink limited during the day. This is important from the point of view of control, because adjusting of environmental conditions in response to the actual state of the plant in stead of according to the average state over a day can improve the efficiency of the process. A comparison of model and data in an optimal control experiment are shown in Figure 2.3. Despite considerable model reduction a description of fruit weight is obtained, which is satisfactory in view of the rather large variance of the measurements. No attempt has been made to use the original, large model, because this would have required an extensive and difficult recalibration.

A completely different approach is to approximate the behavior of the complex model by black box, i.e. by data oriented models. The procedure is to first generate an extensive set of pseudo-data in the range of interest from the complete model. Next, a

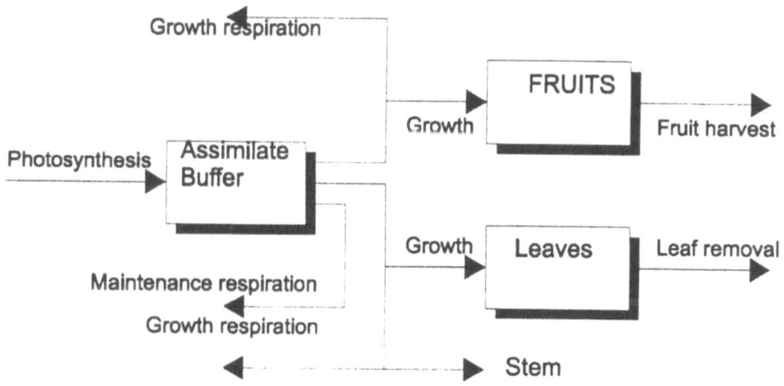

Figure 2.2. Reduced model: Big-leaf-big-fruit tomato model structure (Tap and Van Straten 1995).

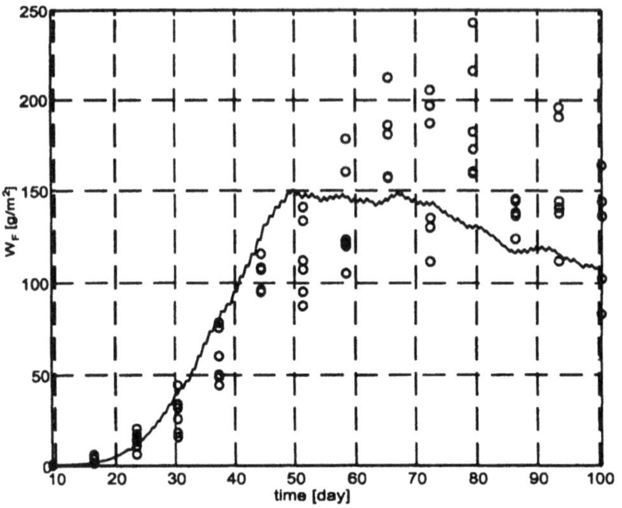

Figure 2.3. Calibration of big-leaf-big-fruit model. Fruit dry weight. Notice the large variation in the data (circles).

black box model is developed from the data by ordinary system identification methods. In this way Young and Lees (1996) approximate the greenhouse physics by a data-based mechanistic model of low order. In order to cope with non-linearities, a non-linear input-output model may be chosen, such as neural nets. Seginer and Ioslovich (1996) showed that TOMGRO containing 69 state variables, could equally be described by a neural net with an equivalent state vector with dimensions one order of magnitude less.

2.4. Validity domain and need for calibration

One may think that in the ideal case parameter estimation is unnecessary. Each model, however, has associated to it a validity domain. Validity has several attributes:

- the amplitude and frequency range of input signals;
- the assumptions about the non-modeled environment of the system;
- the properties and internal structure of the system itself.

More often than not the validity domain of a model is not well documented. Greenhouse crop models usually have been developed for environmental inputs prevailing under 'normal' operation. In a search for optimal control, however, the model may need to be used in an input range not yet observed, for example at lower temperatures than usual. The model is than stretched outside its original validity domain, and re-calibration (on the basis of newly designed experiments) is needed. Most crop models assume that the root environment is non-limiting. When in an actual situation this condition is not fulfilled, model predictions will be erroneous. Some differences in environment, for instance the soil type, that have not been modeled, may be accommodated by recalibrating the model. Finally, a model may be designed for a specific crop or cultivar. When used for another one, re-calibration of some of its parameters is unavoidable.

More often than not the validity domain of a model is not well documented

The need for re-calibration restricts the general applicability of a model. Efforts can be made to collect as many re-calibrations as possible, to create a table from which the likely parameters for various crop/cultivar combinations, under various non-modeled environmental conditions can be read. Such a table is, in fact, a static data-based black box sub-model, mapping application conditions to parameters of the model.

General applicability is not necessary in a control environment. It is quite conceivable to calibrate a model for the particular situation at hand, and use it for subsequent instances of control and operation with the same crop in the same environmental setting.

2.5 The exploitation of data

In the above, mechanistic models have been taken as the starting point. We will now consider the role of data. First, data can be used for model calibration. An obvious requirement is that calibration of the model is, indeed, possible. In complex models, this can be quite problematic, because of structural identifiability problems, and problems of identifiability from the available data (Metselaar 1999). This is another reason to look for simplified model structures. There has been little systematic research, however, to see how parameter identifiability affects the predictive power of models. A parameter that is not uniquely identifiable from the data indicates that output is not sensitive to this parameter. Consequently, its precise knowledge is not required to predict *in the same output domain*. Similarly, correlation between parameter estimates hamper identification of each parameter separately. Cross-correlation, however, leads to lower prediction uncertainty as compared to uncertainty if each parameter was sampled from its own confidence interval.

Until now we assumed that data were already present. The situation is better if we have control over the system. In fact, experiments can be designed to obtain the largest possible information with respect to parameter identifiability. In bioprocess operation, identification-optimal steering inputs have been computed by minimizing the condition number of the Fisher information matrix (Munack 1989):

$$u^{*}(t) = arg\min \frac{\lambda_{\max}(F)}{\lambda_{\min}(F)} \quad with \quad F = \int_{0}^{t_f} (\frac{\partial y}{\partial p})^{T} Q \frac{\partial y}{\partial p} dt$$

Here, y is the model output, u the controllable input, p the model parameters to be identified, Q a weighting matrix, $\lambda_{max}(F)$ and $\lambda_{min}(F)$ the highest and the lowest eigenvalues of the Fisher information matrix F, respectively. Such dynamic designs are not very common in crop growth modeling, but can be well rewarding. The method could be applied as well when pseudo-data generated by a large simulation model are used for calibrating a reduced model (Section 2.1).

In well defined situations, e.g. a specific greenhouse, with a specific crop/cultivar, an attractive alternative to mechanistic modeling is to use the input-output data to generate a black-box dynamical model. A particularly suitable form is given by

$$y_{k+1} = f(y_k, y_{k-1}, \ldots, y_1, u_k, u_{k-1}, \ldots, u_{k-m})$$

This is known as a NARMAX form (Non-linear Auto-Regressive, Moving-Average with eXogenous variables). The function *f(.)* represents a non-linear mapping, and can be conveniently realized by an artificial neural or neural-fuzzy net. Particularly convenient are networks that are linear in the parameters. This can be obtained by taking the non-linear input part randomly as is done by the Random Activation Network RAWN, (Te Braake and Van Straten 1995), or by choosing it on the basis of qualitative knowledge (e.g. the selection of centers and widths of radial basis functions, or of fuzzy membership functions in a fuzzy-neural approach). Figure 2.4 shows the performance of the fuzzy-neural model NUFZY with 4 inputs with 8 inverse multiquadratic membership functions each, and having 10 rules, to describe the growth of tomato (Tien and Van Straten 1995). The number of free parameters is equal to the number of rules in this type of model.

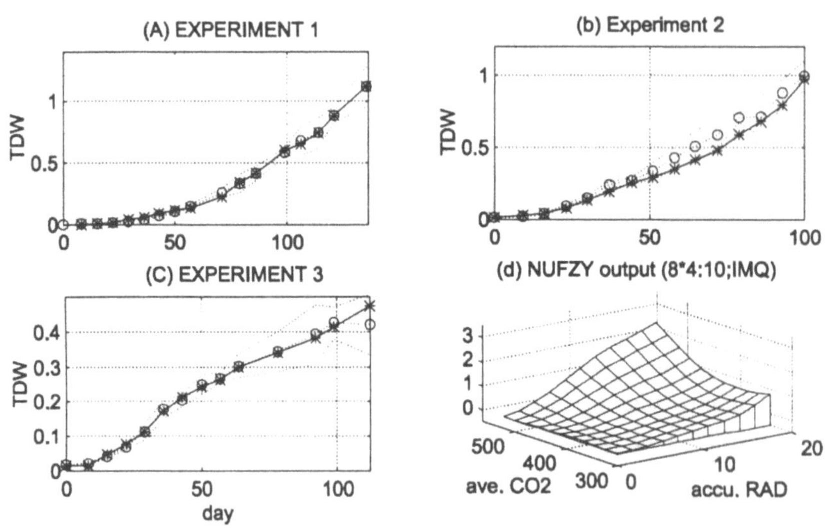

Figure 2.4. The measured TDW of tomato (circle-dotted line) with 95% confidence interval (dotted line). Experiment 1 (a) and 3(c) are used for calibration of NUFZY (star-dotted line), experiment 2 (b) for validation. The predicted TDW by NUFZY is related with accumulated radiation (kW m^{-2}) and the CO_2 concentration (ppm) averaged over the period from starting day until the measurement day.

2.6. Handling mismatch

Even after calibration there is usually mismatch between model and data. The first source of mismatch is in the variance of the data, as shown in the example of Fig. 2.3. This is not a problem of the model, but indicates that real systems are subject to variability. Some of it may be identified. For instance, spatial variability is known to be a major source of variability among plants in a greenhouse. One remedy would be to design and operate a greenhouse such that spatial variability is reduced to a minimum. CFD (Computational Fluid Dynamics) techniques can be used to study the options to accomplish this. The other approach is to model - and perhaps consciously exploit - spatial variability. This would tremendously increase the modeling effort, however. The other source of mismatch is an inherent attribute of modeling. Modeling, by its very nature, means simplification. Some discrepancy, therefore, must be accepted if the model is to serve a useful purpose.

The first task when dealing with model mismatch is to judge the acceptability of the discrepancies. In the ideal case, this should be done with respect to the usefulness for the final goal. For cultivation control, we must check how model errors affect the final net income of the farmer. Research regarding the sensitivity of the final operation and objective function to modeling errors is almost absent, but is urgently needed.

2.6.1Reducing the error 1: Stochastic models.

To judge the effect of modeling error on prediction performance it is necessary to have a description of the error. The most common way to do this is by adding noise terms to the model, as follows:

$$x_{k+1} = f(x_k, u_k, v_k) + G\eta_k$$

$$y_{k+1} = g(x_{k+1}, u_{k+1}, v_{k+1}) + v_k$$

Here G is a weighting matrix, η_k and v_k are random processes, called systems noise and output noise, respectively. The model is stochastic, rather than deterministic, although

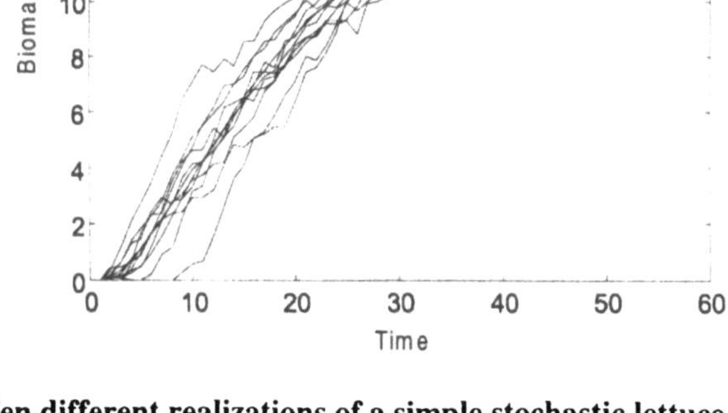

Figure 2.5. Ten different realizations of a simple stochastic lettuce model, illustrating the effect of system error.

deterministic with stochastic terms would be a more appropriate name. The systems noise can be considered as a signal generated by a noise filter driven by a uniform independent stochastic signal. If it is assumed that the driving signal has a normal distribution, then the approach is probabilistic. As an alternative it can be assumed that the driving signal is just unknown but bounded, leading to a set-theoretic setting (e.g. Schweppe 1973). The art of modeling is to deduce distributional properties of the systems noise from data in the presence of measurement noise. This can be done by systems identification techniques (Ljung and Glad 1994).

Predictions made with the model can only be made by generating a realization. A number of realizations yield an output with a mean and a distribution. Figure 2.5 shows the behavior of a stochastic mechanistic lettuce model for ten different realizations of the systems noise in a hypothetical example. Assuming systems noise in stead of just output noise often provides a better description of the expected variability. An indicator for the presence of systems noise is the prediction error sequence, i.e. the sequence that results when predictions are made from previous data. If this sequence still contains signal information, there are most likely modeling errors. The parameter estimates obtained from erroneously assuming an output error structure are biased. Enhancing the systems equations by a stochastic term often dramatically improves the expected prediction error, and reduces the parameter bias.

Enhancing deterministic models with a stochastic part does not reduce uncertainty, but it reduces the bias in parameter estimates and increases the reliability of the forecasts.

2.6.2 Reducing the error 2: Model expansion.

Suppose that by analysis or qualitative judgement the model mismatch is unacceptable. It is then conceivable to expand the model, or to revise the formulation of its sub-processes. This pathway is the choice if it comes to increasing our understanding. Modeling is a powerful tool to organize our thoughts and to guide experimentation. A disadvantage of model expansion is that it increases the complexity, and entails the need for further experimentation, associated with a parameter estimation exercise. If operational control is the purpose, it is desirable to develop methods that will assess the control consequences of an expanded model in order to balance the degree of detail against practical usefulness.

2.6.3 Reducing the error 3: Internal optimization models.

A common source of considerable modeling difficulties in crop models is the need to describe the allocation of resources to different parts of the plant. Similarly, it is often difficult to model preferences of micro-organisms to various substrates in bioprocesses. Biosystems appear to adapt to changed environment, by changing their internal structure, both on the species level and on the community level, to maximize their biomass or chances for survival. In a full mechanistic model it is necessary to model these internal regulation mechanisms, so that optimizing behavior on the macroscopic level is an emergent property. This, however, is an enormous task, and would involve an approach perhaps on the molecular or gen level.

An interesting option to avoid explicit detailed modeling of adjustment mechanisms and self-organization, is to incorporate optimizing behavior into models. In this way internal-optimization models, sometimes called cybernetic models (Kompala et al 1986), are obtained. The procedure is to introduce time-variant parameters that act as control variables into the equations of change, and then compute the time pattern of these control parameters so that a specific goal is optimized over time.

As an example, consider the system

$$\dot{x} = p(t)[f(x) - \alpha \cdot x - \beta \cdot y]$$
$$\dot{y} = (1 - p(t))[f(x) - \alpha \cdot x - \beta \cdot y]$$

where x is the vegetative plant part (leafs), y the generative part (seeds), $f(x)$ the photosynthesis, and $\alpha{\cdot}x$ and $\beta{\cdot}y$ the loss due to maintenance of leafs and seeds, respectively. The optimization criterion is that at the end of the growing season T the seed yield should be optimum. Then, the task is to find the optimal pattern $p^*(t)$ such that $y(T)$ is maximized, or

$$p^*(t) = arg \min\ y(T; p(t))$$

This problem can be solved with optimal control theory. A particular useful method is the optimization with the Hamiltonian function, introducing co-states (Lewis 1986), which represent the marginal value of a unit of leafs or seeds at any time. Fig. 2.6 shows that first all assimilates

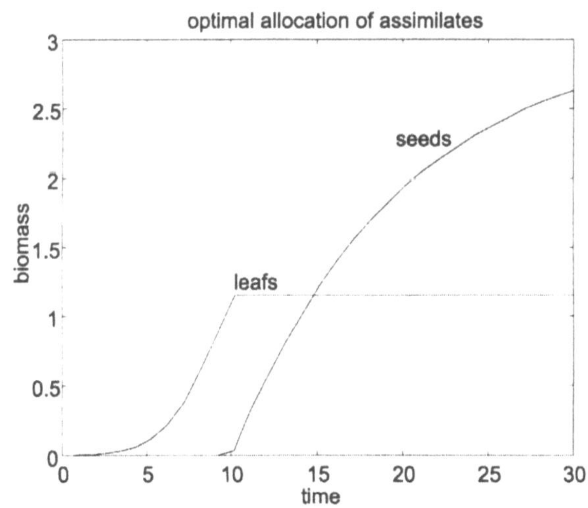

Figure 2.6. Optimal allocation. The allocation function (not shown) switches from 1 to 0 at $t = 10$.

are assigned to the leafs ($p = 1$). Then, after a time determined by the optimal solution, all efforts are put to produce seeds ($p = 0$). This result is achieved without explicitly modeling how p depends upon the system states or the environment. Iwasa and Roughgarden (1984) discuss several more elaborate cases in their seminal paper. Velten and Richter (1995) have shown that maximizing biomass as a goal function yields similar results.

2.7. The introduction of feed-back

With a suitable model, cultivation control can be achieved by computing optimal steering patterns over a season. This is an open loop solution. If the actual weather conditions deviate from the expected, this may become sub-optimal. Also, modeling errors accumulate. The classical solution to abate uncertainty is the introduction of feed-back. In crop cultivation, this would mean the on-line observation of plant behavior. Several approaches have been proposed, e.g. the speaking plant concept (Hashimoto 1989), and the use of image processing techniques (Van Henten and Bontsema 1995). Feed-back together with self-learning models will be a challenging issue for further development of crop cultivation control in the future.

2.8. Summary and Conclusions

Crop cultivation steering requires concise but sufficiently accurate models. Several options to arrive at these control oriented models have been discussed. Reduced mechanistic models can be developed by reasoning, or by formal identification techniques applied to large comprehensive crop growth models. Identification methods can also be applied directly to the data to derive non-linear black-box models, such as neural nets or fuzzy-neural models. Enhancing deterministic models with a stochastic part does not reduce uncertainty, but it reduces the bias in parameter estimates and increases the reliability of the forecasts. Finally, if detailing of models is unavoidable, cybernetic modeling may be an interesting option to solve the problem of resource allocation.

If detailing of models is unavoidable, cybernetic modeling may be an interesting option to solve the problem of resource allocation.

Exercises

1. Fig. 2.1 describes a very general system. Try to formulate the pea model in terms of this particular system. Specify an output variable, the model and various time points.
2. Suppose (in section 2.5) that a model exhibits the following behavior: $y(t_1,t_2) = b \cdot T \cdot (1+\sin(t/a \cdot \pi))$ where T is the temperature and t is the time of the day. Determine $\partial y/\partial T$ and $\partial y/\partial t$. Suppose that both variables are equally important – leading to the matrix Q being the identity matrix. Determine the Fisher Information Matrix F.

Suppose that data are given as $(y,T,t) = (3,5,12),(3.5,7,8),(4,15,8)$ and $(4,20,11)$. Determine the minimum and maximum eigenvalues of F and hence the minimized condition number $u^*(t)$ Interpret your results.

3. Check the validity domain of the pea model, or any other crop growth model you know, according to the criteria of Section 2.4.

4. A very simple model for the photosynthesis sub-process is

$$P = \frac{\varepsilon I \,\sigma\left(C - C^*\right)}{\varepsilon I + \sigma\left(C - C^*\right)}$$

where P is the photosynthesis rate (kg CO_2 m^{-2} s^{-1}), I the photosynthetically active light intensity (W m^{-2}), C the CO_2-concentration in the air (kg CO_2 m^{-3}), with parameters ε (light use efficiency in kg CO_2 J^{-1}), σ (canopy conductance in m s^{-1}) and C^* (CO_2 compensation point). Suppose that ε is known to be 13.10^{-9} and that C^* is 3.10^{-5}. The following table shows triplicate measurements at a fixed CO_2-concentration of $6.875.10^{-4}$ (350 ppmv):

I	P (*10^{-7})	P (*10^{-7})	P (*10^{-7})
30	2.79	3.09	2.96
65	4.31	5.05	4.73
100	6.60	6.90	6.68
128	7.49	7.05	6.68
161	7.50	9.17	8.43
199	9.31	8.65	9.53
230	9.73	9.19	8.79
268	9.53	10.08	9.97
302	10.01	9.88	10.48

a) Calculate the least squares estimate for σ.
b) Calculate the range of σ such that the model curve is outside the upper and lower bound defined by the data at at most one light intensity (unknown-but-bounded estimation).

5. Using the pea model, calculate by simulation the statistical distribution of the final biomass in response to an assumed a priori distribution of a parameter of your choice.

6. Simulate the internal optimization model of section 2.6.3 and try to find the optimal switching time by hand. Model for f:

$$f\{x\} = c_m \frac{x}{K_L + x}.$$

Recommended parameters: $\alpha = 0.01$; $\beta = 0.05$; $c_m = 1$; $K_L = 5$. Initial conditions: $x(0)=0.1$; $y(0)=0$. Maximize seed at final harvest at $t = 100$ time units.

3 Finding and using data for small scale applications of agrometeorological models such as yield forecasting at a European scale

P Vossen

3.1 Introduction

Since 1993, the European Union is operating a system for yield forecasting at the level of the membership countries and of large sub-national regions. Crops of interest are major cereals, oil and protein crops and olive and grapevine. Forecasts are valid at the level of the EU as a whole and of the EU memberstates. The system is operated by the Joint Research Center as part of the Project for the Application of Remote Sensing to Agricultural Statistics created in 1988 by the EU Council of Ministers. The main clients of the Project are the EU's Directorate General for Agriculture and the European Statistical Office (EUROSTAT). Since 1996, a similar system is being developed for the central and eastern European countries and for the Maghreb area.

Application at the scale of large regions and countries of agrometeorological models for yield forecasting, goes along with a number of problems and constraints. These result from non-adequacy in input data requirements between scale of application and local scale at which the models were initially developed. On the basis of the system of the European Union for crop yield forecasting, we present and discuss in this chapter specific aspects related to collection and processing of input data. Examples include meteorological and soils data, but also estimation of derived parameters such as potential evapotranspiration and global solar radiation and statistical validation of model outputs. Special attention is paid to constraints related to introduction of region-specific crop knowledge in a simulation model, interpolation of input data, validation of outputs and limitations in the use of the results. These limitations are such that almost no downscaling of system outputs is permitted and that, what initially starts as a deterministic simulation of crop biomass and grain development, eventually results in indices whose inter-annual variability varies along with the inter-annual variability of grain yield at a regional or national scale, but not necessarily in an absolute simulation of grain and biomass.

The following implementation strategy was opted for:

1. Crop yield forecasting, at least in the initial stages, is mainly to be based on traditional methods like agrometeorological crop growth simulation models, the aeropalynological method and technological time trend extrapolation. Proven methods that relate satellite imagery to quantitative crop yield forecasts at the national or regional scale not yet being available. The applied techniques have indeed shown their applicability at a regional or national scale (Brochet et al 1975; FAO 1986; Haun 1982; Motha et al 1986; Place and Brown 1987; Sakamoto 1978; Vossen 1990a).

2. Simultaneously, possible use of satellite remote sensing techniques for crop monitoring and yield forecasting and for improvement of precision and spatial resolution of the agrometeorological model outputs would be investigated.

This chapter discusses how the previous strategy was followed for finding and using data for quantitative yield forecasting at national scale.

3.2 The models for crop yield forecasting at the scale of the EU

Three type of models have been developed tested and implemented:

- Deterministic crop growth simulation models for annual crops and grassland;
- The aeropalynologic method for grapevine;
- A simple water balance model for the perennial grapevine and olive tree crops;

Each of the models is briefly summarized hereafter.

3.2.1 Deterministic models for annual crops and grassland

For annual crops, deterministic agrometeorological crop growth simulation models each derived from the WOFOST model (Van Keulen and De Wolf 1986; Van Diepen et al 1989; Supit et al. 1994) have been validated for cereals, grain maize, rice, soybean, field bean, potato, sugar beet, oilseed rape and sunflower (Vossen 1990, 1992). The models are driven by a combined energy balance - water balance module. This compares real transpiration with calculated potential transpiration through a sub-module combining light interception, CO_2-assimilation, water requirements and water availability. Models are calibrated to make them region-specific. Initial dry matter at emergence, mean planting date, calibration of the length of phenological stages as a function of sums of temperatures are determined on the basis of site specific field data available from various research institutes in the EU. Also crop knowledge bases have been applied which were specially established for this purpose (see the annex to this chapter). Both adaptation and calibration have been reported on in Boons-Prins et al (1993).

Areal monitoring of the state of grassland at the level of the EU was derived from two existing models. For the northern part of the EU, the LINTUL (Spitters 1987; Kooman 1995) was adapted and validated. This model applies to grasslands with winter dormancy and restart of the vegetative growth after winter. To simulate biomass production, exploitation of grass by cattle or farmer is taken into account, as well as the fact that the grassland almost never flowers. For the southern parts of the EU, with mainly annual grasses, the ARID CROP model (Van Keulen 1975) was validated. Grass cover, composed of a mixture of annual grasses, generally passes through a global cycle emergence, vegetative growth, reproduction, ripening and death. The software module which integrates those two approaches is referred to as the LINGRA Model (Bouman et al 1996.).

3.2.2 Simple water balance models for the perennial grapevine and olive tree crops

For most perennials in the EC, deterministic growth simulation models are not available and their environmental growing conditions, phenological calendar, and farming practices (including the main cultivated varieties) have, in most cases, only partly be described in the literature. The same applies for the soil types on which the various perennials are grown and for the rooting depths which are highly variable according to the general climate, soil type and farming type (irrigation, for example). It is thus difficult to propose for perennial crops of major economic importance such as olive tree and grapevine, a simulation model that needs -as an input or for model validation- information such as sprouting period, available water capacity, period of flowering, etc.

A simple water balance model for monitoring and forecasting of the national yields of olive oil and grapevine had therefore also to be developed. This OLIWIN model (Bories 1996) is largely based on Vossen (in: Riou 1993; Vossen and Rijks 1995), Doorenbos and Kassam (1979) and Doorenbos and Pruit (1984). In its present state, the model estimates the soil moisture content and its variations on a monthly basis from April to September, which are the average sprouting and ripening months within the EU. To estimate the water requirements and water consumption, a crop factor K_c, expressing the ratio between water requirements during a development stage and potential evapotranspiration, and an estimate for the profile available water capacity are used (King and Le Bas 1994). Soil moisture at the end of March is estimated from the October-March rainfall. However, additional data collection is ongoing, to make the model more region- and soils specific, and to take into account variable dates of sprouting and flowering.

3.2.3 The European Commission's Crop Growth Monitoring System

The 3 models WOFOST, LINGRA and OLIWIN are integrated in the EC's Crop Growth Monitoring System (Vossen 1990, 1992; Vossen and Rijks 1995). This system has been designed to use daily weather data (rainfall, temperature, vapor pressure, 24-hour wind run, sunshine duration or cloud cover) that are interpolated to a regular grid covering the whole of Europe. It permits to run the various models for different soil types on which a crop is cultivated, taking into account region-specific average planting or sprouting dates and crop parameters. For yield forecasting at the national scale, model outputs are weighted according to the relative importance of soil type and crop acreage in the various regions of a country. Validation for forecasting is done once every month, between the month of expected planting and the harvest month. In this procedure, time trend and model outputs at the end of a given month, are simultaneously regressed against historical yield series. This approach assumes that the weather conditions during the remaining part of the season will be "normal". The approach also implies that the forecasts become more and more reliable according as the season progresses.

The following outputs are available as tables or cartographic products a few days after data acquisition: biomass and grain production, under the actual rainfall conditions and as if all required moisture were available; estimated actual soil moisture reserve and state of advancement of the cycle; derived meteorological parameters such as sums of temperature, climatic water balance, etc.; percentage departure from the long term mean given decade or period within the growing season, for all of the previous indicators.

3.2.4 The aeropalynologic method for olive tree and grapevine

Simultaneously, but only for grapevine, the aeropalynologic method for the forecasting at flowering of potential grapevine production was further validated and implemented at the scale of wine producing regions of the EU. This POLLEN method (Besselat and Cour 1990) is based on the assumption that the number of pollen liberated into the atmosphere during the flowering period of certain crops. It is a good indicator of potential crop yield. In fact, intensity of pollination may in some cases integrate growing conditions previous to flowering. Further details on this method and its results are provided in Besselat and Cour (1996).

3.3 Inventory of problems related to the small scale application within the EU context of models for yield forecasting.

Implementation of a system for production of timely yield forecasts at a national or regional scale, goes along with the following problems:

1. A change of scale, from site-specific input information, to assessments or forecasts that are valid for large regions or countries;
2. A limited precision of the input information. For example, the information contained in the EC's 1:1.000.000 soil map (CEC 1985) is already a generalized synthesis of the original field observations. Meteorological conditions at the level of weather stations are not always comparable with the weather conditions in the farming areas, nor with weather conditions recorded at the research sites were the relations between crop growth and weather conditions were established. Planting dates and crop cycle lengths can only be estimated with a precision of approximately 5 to 10 days. Part of the available input information is only valid within national boundaries, such as soil classification criteria and type and installation characteristics of meteorological instruments).
3. Non-availability of part of input information. Examples are the exact profile available water capacity of the soil (PAWC), depth and rooting depth of the soils, soil coverage by various crops in a region, relative importance of various crop varieties that are planted in a given region.
4. A limited spatial resolution of the input information. The number of reliable, daily reporting weather stations in the EC is limited to approximately 650. A complete and reliable time series of crop yield, needed for the validation of any model or method, are only available for the EU. For crop growth conditions, farming practices, etc., information is in most cases only available for regions exceeding 1000 km^2.

5. Knowledge is often only available at the level of individual research sites on the relations between crop yield and the agro-pedo-meteorological growth conditions. It is never integrated into information valid at the regional scale.
6. Non timely availability of the information on farming practices and farmers decision making mechanisms, for example the planting dates during a given season, the fertilizer doses, etc.

From what precedes, two conclusions can be drawn. First, quantitative yield forecasts can never be valid for a specific locality but only for (very) large areas like countries or regions. Information and model outputs should have been carefully weighted for their relative importance, within the large region, of soil types, groups of varieties, common farming practices, etc. Secondly, yield forecasting has to be validated per country and/or per large region. These validations have to be done with statistical techniques (regression, analysis of variance), to quantify effects such as technological trend and to take into account the variable impact of environmental conditions according to farming level and farming conditions.

3.4 Activities undertaken to solve part of the listed problems

The above listed problems and constraints mainly relate to the non adequacy in input data requirements between the scale of application and the local scale for which the models were initially developed. As a partial remediation, a number of applied research activities was undertaken.

3.4.1 Establishment of agro-pedo-meteorological crop inventories.

For most of the crops commonly cultivated in the EU information was collected on (1) statistics on cultivated surfaces and yields, (2) regionalized phenological crop calendars and (3) environmental requirements of crops. The list of available crop inventories is given in the appendix to this chapter. Each inventory contains:

a. General crop data such as rooting depth, base temperature, crop coefficients expressing the ratio between crop waiter requirements and potential evapotranspiration during a given development stage, etc. This information was mainly obtained through literature searches.
b. Agronomic data, such as the earliest and latest dates of harvest in a region, the most common farming practices, the major varieties grown in a region, maximum altitude at which a crop is grown, technological evolution over the last years, etc. This information was mainly obtained from postal questionnaires.
c. Physiological information from individual trials, such as the duration of phenological phases. This information was also mainly obtained through literature searches.

Limitations of the knowledge bases (Russell 1990). Information is often available at the wrong scale. For example, sowing dates are often only known for a small sample of fields, which may or may not have been biased. Information about the representativity of the subsamples, needed for accurate scaling up from the side-specific information to a regional or national level is usually unavailable. Similarly, crop modeling parameter values obtained from individual trials will differ from those that would have been obtained if a complete enumeration had been achieved. This is of particular importance if relations are not linear (De Wit and Van Keulen 1987). Also, certain information is frequently unavailable for certain parameters and in certain regions or countries. In such cases a 'best guess' needs to be made.

Information is often available at the wrong scale

Both time and space resolution of the outputs will be limited. First, it is not reruired to exploit daily model outputs. Estimates of dates of occurrence of phenological stages, are most likely imprecise and probably only suitable to be expressed in relative terms such as departures from normal or "very late", "late", "normal", etc. Second, using mapped crop state monitoring products for spot evaluations is not permitted. Outputs should not be presented as isolines, but assigned to grid cells (for qualitative monitoring) or as tables with summaries at the national scale (for quantitative yield forecasts). In practice a resolution of 50×50 km^2 is used which corresponds also to the resolution to which daily meteorological data, available in real time from the European network, can be interpolated with confidence (see further).

3.4.2 Estimation of solar radiation and potential evapotranspiration

Both PET and solar radiation are key input variables in many crop yield forecasting models. Regional validity of most estimation methods, however, is in most cases limited. Methods for calculation of both parameters have been developed by Jones et al (1991), Choisnel et al (1992) and Supit (1994). Their approaches could only be validated on the basis of limited sets of observations: for solar radiation on 100 stations in the EU, Central and Eastern Europe, the Near East and North Africa countries and for PET on 6 sites. Therefore, large areas lack sufficient data points, leading to rather imprecise isolines. Given the scarcity of literature in this fields, however, the here presented results are the presently the best to be used.

For estimation of global radiation from visual observed cloud cover at meteorological station level three equations are being used, depending upon available surface meteorological data:

a) If daily sunshine duration (n, in hours) is available (Ångström 1924; preference choice):

$$R_{g,d} = R_{o,d}(c_a + c_b \frac{n}{D}) \qquad (1)$$

b) If both cloud cover (*cloud*) and maximum and minimum temperature T_{max} and T_{min} are available:

$$R_{g,d} = R_{o,d}\left\{c_a\sqrt{T_{max} - T_{min}} + c_b\sqrt{1 - cloud/8}\right\} + c_c \qquad (2)$$

c) If only temperature data are available (Hargreaves et al1985):

$$R_{g,d} = c_a R_{o,d}\sqrt{T_{max} - T_{min}} + c_b \qquad (3)$$

where $R_{g,d}$ is the observed daily global radiation ($MJ.m^{-2}.d^{-1}$), $R_{o,d}$ are daily values of the extra terrestrial radiation ($MJ.m^{-2}.d^{-1}$), D is the astronomical day length (h) and c_a and c_b are empirical constants.

The above 3 approaches were tested for more than 100 stations ranging from Northern Finland to Southern Egypt. Generally the Ångström method provides the best estimates. However Hargreaves et al (1985) is an acceptable alternative when no daily global radiation and sunshine duration observations are available, which is often the case for daily reporting meteorological stations.

For potential evapotranspiration the original Penman formula (1948), is being used, but by using any of the above methods for estimation of solar radiation (Choisnel et al 1992). These modifications give acceptable estimates of PET over 10-day periods between March and October. During this period, the mean deviation between daily lysimeter observation and estimated PET, the mean of the absolute deviations and the mean quadratic deviation are 0.1, 0.55 and 0.71 $mm{\cdot}d^{-1}$, respectively, as an average of mean values at the 6 lysimeter stations. Although the Penman-Monteith approach gives comparable results in terms of difference between observed and estimated PET, it has not been implemented in the system because part of the necessary input information (crop canopy resistance) can not (yet) be made available for use on a regional scale as the system requires.

3.4.3 Soil data

For use at the European scale as a whole, only the existing 1:1.000.000 EC soils data base (CEC 1985, FAO classification) is available. This data base was initially a compilation of national maps, causing problems such as lack of cross-boundary harmonization, differences in classifications, errors in digitalization and map attributes, different levels of precision in description and aggregation of soil types into soil mapping units, etc. The soils data base was therefore corrected, harmonized, updated and completed with information available in the original archives. This eventually resulted in Version 3.1 of the EC soil map (King et al 1993). The soils data base as such is mainly used in conjunction with the crop knowledge bases, to identify the areas where a given crop can possibly grow. From it, a Profile Available Water Capacity Map has been derived (King et al 1993). The latter data base contains for each soil typological unit (STU) in a given soil mapping unit (SMU):

Figure 3.1 The network of actually available meteorological stations, providing daily data in real time.

- a PAWC estimate for the first 3 main horizons (or less if an non penetrable horizon is present) obtained from a set of pedotransfer rules;
- an estimated maximum possible rooting depth;
- an indication whether moisture could become available through capillary rise;
- the importance of the STU expressed as a percentage of the total area of the SMU.

3.4.4 Meteorological data and their processing.

The number of stations for which actual daily data of rainfall, temperature, vapor pressure, wind speed and cloud cover or sunshine duration are available for the EU is approximately 650. For only 350 stations, however, long term historical series are available. These networks are relatively sparse considering the large range of relief and elevation. At present no proven techniques exist for interpolation of daily meteorological data from the synoptic network and applicable to the EC as a whole. Such techniques are only available for a limited number of variables and only for a few countries. Therefore, a new technique was developed for interpolation of data from the existing network of meteorological stations on a regular grid. (Van der Voet et al 1993).

The technique can be summarized as follows: values are estimated by inverse distance weighting of the data of an optimum set of stations, surrounding the center of a the cell of a regular grid mesh over Europe. A set of surrounding stations is selected on the basis of criteria of similarity between station characteristics and the grid cell to which is being interpolated. Criteria of similarity are: proximity, similarity in terms of altitude, distance to the coast, position relative to climatic barriers such as Alps and Pyrenees, degree to which selected stations surround the grid cell and the number of stations in a set. Initially, the 7 nearest stations surrounding a grid cell are selected. For each combination of 4 of these 7, a 'suitability score' is calculated on the basis of empirically validated weights, that are attributed to the values taken by the similarity criteria. An interpolation to the center of the grid cell is then made from those stations. Increasing the number of chosen stations beyond 4 did significantly increase the calculation time but not the quality of the interpolations.

For rainfall, the single station with the lowest difference score is selected. Use of more stations reduced the mean prediction error, but also considerably overestimated the number of wet days, as such modifying the temporal distribution of rainfall, which is of major importance for crop modeling.

According to the results of the analysis of the errors of estimate, the meteorological conditions that are relevant for crop growth, are fairly accurately estimated at a spatial resolution of 50×50 km^2.

Figs. 3.1 and 3.2 give the network of meteorological stations providing daily data in real time and the 50×50 km^2 grid for the EU to which both historic and actual data are interpolated. Only those stations are represented from which input data required for running the models on a daily basis can be obtained. In practice it represents the SYNOP network with at least 4 available observations per day. The list was submitted to an expert group to screen those stations neither be used for agrometeorological crop state monitoring nor for alarm warning, because of their location. For example, stations located on a small island or at the edge of the coastline, surrounded by the sea, in the center of a very large city or at a close proximity to an airport runway were excluded. To identify the stations that can be used, the data from all available SYNOP stations are once a year submitted to quality tests (Meteo Consult 1991) and the stations that reported too irregularly or too unreliably during the previous year are excluded for use during an actual season.

The number of stations for which actual data are available is far higher than the number of stations for which long term historical series are available. This is partly compensated by running the crop model and calculating derived parameters such as sum of temperatures, climatic water balance, at the level of each grid cell, both for historical and for actual data. This is necessary, because most relations between plant growth and agrometeorological growing conditions, and the spatial evolution of many meteorological and pedological conditions are often non-linear (De Wit and Van Keulen 1987).

From what precedes it follows again, that yield forecasting methods using meteorological information can not give reliable outputs at the local scale and that the agrometeorological data or model outputs obtained for a given station or grid cell, can only serve as areawise or regional indicators of the quality of a cropping season. To use the information for quantitative yield forecasts, it is necessary to calculate regional averages of local outputs, that are preferably weighted for the area occupied by the crop in a region and the occupation of the different soil types by a crop.

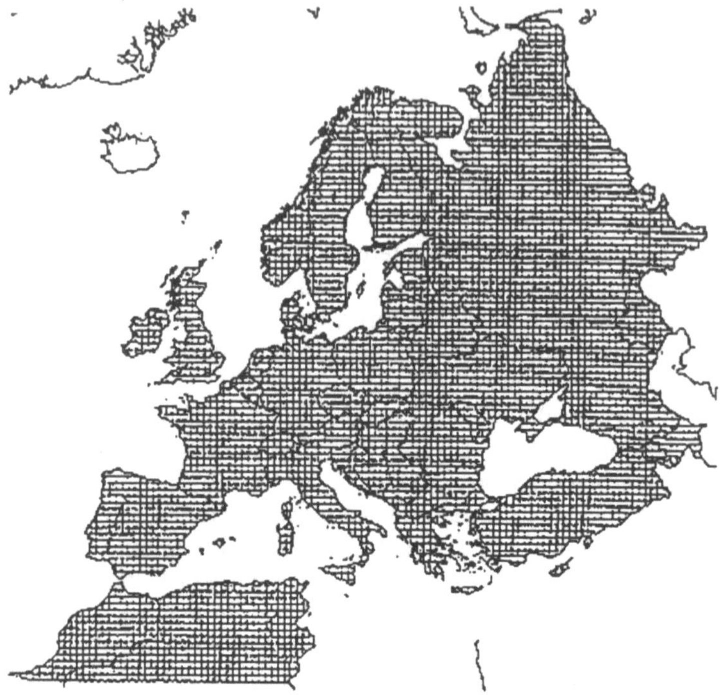

Figure 3.2 The 50 × 50 km² grid over the E.U. to which meteorological data are interpolated.

3.4.5 Time trend analysis.

Crop yields largely depend upon farming practices. These may vary more or less rapidly, according to techniques that farmers adopt and their flexibility to introduce innovations. However, this information is not timely available for yield forecasting purposes at a national scale. Crop growth modeling thus assumes stable environmental growing conditions (except weather) that do not change from year to year. This assumption is obviously not true for farming practices such as fertilizer application and variety choice. On the other hand, a time trend of regional and national crop yields, may to some extent replace farming conditions that affect crop yield but that are not easily available at the time the crop yield forecasts are made. Use of a time trend may compensate changing farming practices from one site to another within the area, for a given year. It may result in a stable general situation for the region as a whole. As a consequence yield predictions can only be valid for areas for which the trend (existing or not) was quantified.

The form of the trend differs between crops, regions and, often, periods within a given time series. Correct description of time trends is important for crop yield forecasting. Palm and Dagnelie (1993) carried out a comparative study for most crops in different member states. Their main conclusion was that, although more complicated models result in a (very) limited gain of precision, the technological trend was best approached by the quadratic function:

$$Y = a + b \cdot T + c \cdot T^2 \tag{4}$$

Where the coefficients *a*, *b* and *c* are estimated by stepwise regression.

3.4.6 Validation of the model outputs.

The general procedure for validation of the agromet models has been reported by Vossen (1990, 1992). It basically is a regression analysis of the series of regional and national yields against functions Ψ of model outputs and Θ of (possible) time trends:

$$Y = C + \Psi\{\, Crop\ Model\ Output\,\} + \Theta\{Trend\} + error \tag{5}$$

The possible time trend and model outputs are simultaneously taken into account and validated on the basis of the hypothesis that weather conditions can suppress or modify the expression of a time trend (Nyankori 1979; Vossen 1990a). The tested trend function is of the form $\pm\, b \cdot T \pm c \cdot T^2$, with *b* and/or *c* equal to zero. Usually, analyses including time trends are carried out for time series of 20-30 years. Because of the recent agricultural policy of the EC, aiming at a reduction of production of and/or subsidies for some crops (wheat), the length of the time series used for the statistical model validation has been reduced to 9 years. This length is sufficient to give enough degrees of freedom in a regression analysis with at least 2 and possibly up to 5 independent variables.

The variables are:

(a) the trend;
(b) one of the independent outputs: biomass, grain, leaf area index or development stage;
(c) one and possibly more of the variables: ratio between theoretical water requirements and real water consumption, rainfall in excess to the water requirements (e.g., during a given development stage), climatic water balance during winter (rainfall minus P.E.T). The possible number of variables here is almost unlimited and selection must reflect real constraints to crop production.

At present, only the variables (a) and (b) have been validated. The statistically most significant model output is selected for forecasting. Model precision and stability and trustworthiness of forecasts are evaluated with independent estimates, based on the model validation for the previous 9 years.

The major consequence is that simulation model outputs are not anymore reflecting a reality such as a truly simulated biomass, but have become indicators. Their inter-annual variability possibly accounts for a more or less important part of the inter-annual variability of national or regional crop yields. These indicators still have a biological basis, because of constraints linked to data availability and imposed simplifications and work approaches, but are not anymore an exact representation of a field reality.

3.5 Results

Table 3.1 (from: Genovese 1997) compares the May-September yield forecasts at the level of the EU and for the years 1993-1996, with the estimates available from the European Statistical office. It appears that the May-September yield predictions are in most cases more reliable and earlier available than the Eurostat figures which are mainly derived from time trend analyses completed by inputs provided by national statistical offices.

Further analysis (Vossen and Rijks 1995; Genovese 1997) however, showed that the time trend component is often more significant than the agrometeorological model outputs. At the national scale, equation 2.5, including the time trend component, significantly (1% level) accounts for inter-annual yield variability for 77% of the crop-member state combinations. In only 53% of the combinations, crop simulation model outputs significantly improve upon the result. This does not necessarily mean

Crop	year	May	June	July	August	Sept.	Final
	1993 (E)	5.60	5.46	5.47	5.46	5.21	5.32
	1993 (M)	5.30	5.21	5.18	5.21	5.20	
	1994 (E)	5.45	5.38	5.55	5.57	5.47	5.42
Wheat	1994 (M)	5.23	5.26	5.22	5.18	5.18	
	1995 (E)	5.68	5.73	5.33	5.35	n.a.	5.31
	1995 (M)	5.39	5.38	5.34	5.35	5.35	
	1996 (E)	n.a.	n.a	n.a.	n.a.	5.84	5.88
	1996 (M)	5.50	5.41	5.37	5.47	5.63	
	1993 (E)	4.10	3.96	3.97	4.14	4.17	4.21
	1993 (M)	4.10	4.17	4.14	4.06	4.07	
	1994 (E)	3.88	3.93	3.95	4.04	4.05	4.01
Barley	1994 (M)	4.00	4.10	4.00	4.00	4.00	
	1995 (E)	4.47	4.09	3.80	3.83	n.a.	3.98
	1995 (M)	4.00	3.90	3.90	3.90	3.90	
	1996 (E)	n.a.	n.a.	n.a.	n.a.	4.62	4.63
	1996 (M)	4.16	4.16	4.18	4.29	4.47	
	1993 (E)	7.70	7.68	7.68	7.53	7.84	7.96
	1993 (M)	7.20	7.17	7.16	7.55	7.58	
	1994 (E)	7.87	7.67	7.59	7.42	7.86	7.69
Maize	1994 (M)	7.80	7.80	7.60	7.60	7.50	
	1995 (E)	8.05	8.09	7.88	7.53	n.a.	7.90
	1995 (M)	n.a.	7.60	7.70	7.50	7.50	
	1996 (E)	n.a.	n.a.	n.a.	n.a.	8.30	8.32
	1996 (M)	7.70	7.76	7.80	7.85	8.22	

Table 3.1: Summary of the 1993-1996 monthly yield predictions established by the MARS Project (M) and compared with the predictions available from Eurostat (E). Yield figures are given in tons per hectare. (Genovese 1997).

The time trend component is often more significant than the agrometeorological model outputs

that in all those cases prediction errors also significantly decrease. In almost all cases, however, the system contributes to improved monitoring and yield forecasting of crops, especially at the national scale, as compared to the use of time trend extrapolations alone or assessments based on a subjective interpretation of general weather conditions.

3.6 Discussion

The relative weak contribution of agrometeorological model outputs to the quality of the yield forecasts is somewhat disappointing. The reasons are probably double:

a) The time trend function is in reality not a single smooth line of increasing (decreasing) yields corresponding to gradually but steadily improving (deteriorating) farming practices. In fact, the mathematical expression of the trend should represent the average effect of improvements as irregular jumps. This may mask effects in regression analysis of agrometeorological growing conditions. For example, a strong positive technological trend, favorable agrometeorological growing conditions during a given season, and an unchanged level of farming practices, may result in the "artifact" that time trend extrapolation appears to be a better yield predictor. Moreover, as a result of advanced farming practices and increasing capacity of farmers to adjust their crop husbandry to actual weather conditions, inter-annual variability of yield data is already relatively limited. The following percentage coefficients of variation can for example be given for the 1975-1989 wheat yields:

Germany	14.7	France	16.1
Italy	7.24	The Netherlands	14.2
Belgium	18.6	Luxembourg	21.1
United Kingdom	17.3	Ireland	21.0
Denmark	14.8	Greece	12.7
Spain	24.9	Portugal	26.3

b) Shortcomings at present of the crop simulation model and the system as a whole, for example:

- the one-layer water balance module does not take into account possible capillary rise of soil moisture which can become available to a crop.
- planting date is fixed and has been put equal to the inter-annual mean value;

- interpolation falls short as concerns estimation of rainfall and radiation from sunshine duration or visual cloud cover observations at station level;
- at most two variables are taken into account: one trend variable and one model output;
- errors in the model parameters such as the initial dry matter at emergence, the sum of temperatures to reach a phenological stage and repartition of biomass towards roots leaves, stems, grain.

c) Quality of the yield data used for validation. Precision of the data is, according to specialists of the European Statistical Office, of the order of ± 5% in the northern EC countries and less in the Mediterranean countries. Moreover, precision and reliability of national and regional statistics vary according to whether they are aimed to provide a good estimate at the regional level (e.g., Germany) or at the national level (e.g., Italy).

3.7 Perspective: the use of remotely sensed information.

At present, the entire system is based on information generated from surface observations. This has as two major disadvantages. First, input data have to be obtained from the relatively sparse European meteorological network, which does not allow to retrieve reliable weather information for areas smaller than approximately 50 × 50 km^2. Second, model outputs for each 50 × 50 km^2 grid cell have a limited precision because ground observed information cannot be made timely available for the whole of the E.U. As a result, quantitative yield forecasts can only become reliable when they are applied to large regions or countries. This approach was initially (in 1988) chosen because proven techniques that operationally relate remotely sensed information to national yields were unavailable, at least within the EU with its typical agricultural pattern.

Introduction of remotely sensed information into models will significantly improve spatial representativity and trustworthiness of both model inputs and outputs.

Introduction of remotely sensed information into models will significantly improve spatial representativity and trustworthiness of both model inputs and outputs. Such improvements are expected from appropriate use of VEGETATION, METEOSAT and/or AVHRR information, especially for the following applications (see also: Loudjani 1995):

a. spatialization of inputs such as meteorological data and of outputs such as drought severity indicators;
b. introduction of land utilization information into the system;

c. improved fitting of crop cycle lengths and major phenological stages, for example by improved assessment of the start of the development of vegetation after winter.
d. improved, more reliable and spatially correct depiction of alarm situations such as droughts, extreme colds and abnormal high temperatures;
e. direct input of estimated global solar radiation from METEOSAT imagery and of remotely sensed surface temperatures and evaporation.

3.8 Annex : Crop knowledge bases

Bignon J 1990 Agrométéorologie et physiologie du mais grain dans la Communauté Européenne. Publ. N° EUR 13041 of the Office for Official Publications of the EC, Luxembourg

Carbonneau A, Riou C, Guyon D, Riom J 1992 Agrométéorologie de la vigne en France. Publ. EUR 13911 of the Office for Official Publications of the EC, Luxembourg

Falisse A 1992 Aspects Agrométéorologiques du Développement des Cultures dans le BENELUX et les Régions Voisines. Publ. EUR 13910 of the Office for Official Publications of the EC, Luxembourg

Hough MN 1990 Agrometeorological Aspects of Crops in the United Kingdom and Ireland. Publ. EUR 13039 of the Office for Official Publications of the EC, Luxembourg

MacKerron DKL 1992 Agrometeorological aspects of forecasting yields of potato within the E.C. Publ. EUR 13909 of the Office for Official Publications of the EC, Luxembourg

Narciso G, Ragni P, Venturi A 1992 Agrometeorological aspects of Crops in Italy, Spain and Greece. Publ. EUR 14124 of the Office for Official Publications of the EC, Luxembourg

Riou C (Ed) 1994 The effect of climate on grape ripening: application to the zoning of sugar content in the European Community. Publ. EUR 15863 of the Office for Official Publications of the EC, Luxembourg

Russell G 1990 Barley knowledge base. Publ. EUR 13040 of the Office for Official Publications of the EC, Luxembourg

Russell G, Wilson G 1994 Wheat knowledge base. Publ. EUR 15789 of the Office for Official Publications of the EC, Luxembourg

Exercises

1. Give a statistical interpretation to change of scale. Consider both the arithmetic aspects and the contents. Focus on measurements on meteorological measurements.
2. Discuss the sincereness of the limited precision and non-availability of input information. What are possible remedies for these problems?
3. Discuss the sincereness of spatial variation and lack of sufficient spatial data.
4. Discuss the sincereness of non-timely available information on farming processes. Are there solutions to this problem?

5. Describe in a mathematical form the interpolation procedure developed by Van der Voet.
6. Discuss the problems at the end of section 3.4: describe two studies, one in which model outputs must reflect the reality, and another where this does not matter so much.
7. Table 3.1 could be analyzed statistically, e.g. by an analysis of variance. Describe the use of an ANOVA for this analysis. What are the factors to include? How would you treat the yield predictions for the different months?

II Some statistical aspects of modeling.

Statistical aspects of modeling for production ecology traditionally deal with a limited but important set of activities. In the first place, models have to be *developed*. This is often a disciplinary activity, but statistical models are usually at the basis of observed and modeled relationships. In the second place, models have to be *validated*. Parameters have to be estimated such that the model applies for the particular range of applications that it has been designed for. Also, some models need to be *calibrated*, i.e. they have to be compared with actual measurements. Both require that reliable data have to be collected, preferably using a statistical experimental design or a sampling scheme that gives representative data. In a range of studies also *sensitivity analysis* is of importance. This could lead to a better insight into the parameters that are the most important and influential ones in the particular application. Next, the model has to be tuned to the scale to which it should be applied. This quite often requires *upscaling*. Data have to be summarized, and possibly new classes have to be defined with new values on old variables.

In the chapters that follow two important aspects of statistical aspects of modeling are treated. The first chapter focuses on model calibration, the second chapter on a statistical analysis of scale.

4 Data use and Bayesian statistics for model calibration

MJW Jansen

4.1 Introduction

Before a model can be applied to a specific system, one has to adapt the model's parameters in order to increase the resemblance between the model and the system studied. This tuning of a model's parameters is called calibration. Only after calibration (and validation) a model may be used to predict the behaviour of the system under various scenario's and to support decisions about the system. The decisions may be tactical, as in greenhouse climate control, or strategic, as in land planning.

For simple models, with a small number of parameters, classical statistical estimation theory may provide a solid base for model calibration. But during the calibration of complex, parameter-rich models, of the kind often used in decision support, it may appear that the system observations by themselves are not sufficiently informative. The problem is that the exogenous circumstances under which the data originated constitute only a small part of what the system may experience, so that different parameter settings may result in nearly equal fits to the observations, but to widely different predictions under new circumstances. In such cases one needs to use all possible additional information about the parameters that is available. It would be sensible and economical to remedy data shortage by scrutiny of the literature for additional information on the parameters. Literature data often originated under a broader range of exogenous circumstances, but they rarely pertain to the specific system for which the parameters have to be set. Most often, literature information alone is insufficient to parameterize a model for a specific situation (Kabat et al 1995; Metselaar 1999). When both sources of information are insufficient by themselves, one obviously should try to combine them.

Bayesian statistics, named after Thomas Bayes (Bayes 1764), provides a theoretically simple way to combine existing information about the parameters with information from new data. This feature makes Bayesian statistics particularly attractive for the combination of the diverse types of information required to calibrate a parameter-rich model. Bayesian statistics summarizes knowledge about a model's parameters in the form of a probability distribution. By this fact, Bayesian methods provide an excellent starting point for uncertainty analysis and decision support.

Bayesian methods provide an excellent starting point for uncertainty analysis and decision support.

Probabilities that summarize one's knowledge are called 'epistemic' probabilities. Their status differs from frequentist probabilities, as occur for instance in coin tossing: such probabilities can be measured, whereas epistemic probabilities are more subjective (Box and Tiao 1973; Kroese et al 1995).

The Bayesian viewpoint provides theoretical insights into the many problems one may encounter during model calibration. Moreover, recent developments in Bayesian numerical methods hold a promise for actually performing calibration.

This paper discusses Bayesian methods for calibration and validation of a model. Moreover, it is briefly indicated that these methods connect quite well to uncertainty analysis and decision support.

Section 4.2 describes common problems with the classical analysis of observational data when the number of parameters to be estimated is large. We present the elements of Bayesian statistics in Section 4.3. That section also treats the construction of a distribution describing the information about parameter values retrieved from literature. Section 4.4 discusses how parameter uncertainty is translated into prediction uncertainty. Bayesian calibration forms the subject of Section 4.5. It results in probabilistic predictions. Section 4.6 briefly discusses possibilities to validate such predictions. The consequences for decision support, of having probabilistic predictions, are touched upon in Section 4.7. The paper ends with a discussion in Section 4.8.

4.2 Problems with the classical analysis of observational data

A typical example of a classical statistical problem is the following. Let vector y denote observations of a system, vector θ model-parameters to be estimated, and vector x exogenous circumstances. Vector $f(x,\theta)$ denotes the model predictions to the observation vector y:

$$y = f(x,\theta) + \varepsilon$$

where vector ε denotes the difference between observations and predictions, caused by measurement and model errors. We take as an example in this chapter the above-ground dry matter yield of maize (y) at harvest in Wageningen, that may depend with some model f on the precipitation x through a parameter set θ. Modeling involves an error ε. We assume that ε follows a distribution $g(\varepsilon)$ that may involve a small number of additional parameters. Typical regularity conditions for the above problem are the following:

- *Identifiability*. When the vector $\theta^{(1)}$ is unequal to $\theta^{(2)}$, the vector $f(x,\theta^{(1)})$ is unequal to $f(x,\theta^{(2)})$, i.e. for different parameter sets we obtain different yield calculations.
- *Differentiability*: the vector $f(x,\theta)$ of model predictions has finite partial derivatives with respect to θ, up to order 3; we only consider regular models, with nice, smooth functions.
- *Presence of a randomness model*: the vector ε is random, with known probability density $g(\varepsilon)$. Obviously, one needs some assumptions about the differences between observations and simulations to draw inferences on θ:

There are other regularity conditions, but these three suffice for the present purposes.

Under the above circumstances it is often possible to show that, with a large set of data, an estimator of θ exists that is approximately normally distributed with mean

θ, and with a covariance matrix that can also be estimated (Cox and Hinkley 1974). One obtains a simple approximation of the distribution of the estimator of θ, which happens to be the normal distribution. But the specific form of the distribution is of minor importance, as long as it is known. The problem with calibrating parameter-rich models is that often none of the above regularity conditions is satisfied:

- *Identifiability*. Model predictions can be insensitive to changes of an inactive parameter belonging to a sub-process that is not activated under the prevailing exogenous circumstances. Anyhow, if the number of parameters exceeds the number of observations, the identifiability condition will not be met.
- *Differentiability*. Many parameter-rich models respond discontinuously to parameter changes, hence violating the differentiability condition (Metselaar 1999).
- *Presence of a randomness model*. The observations of a system can consist of disparate quantities, e.g. time series of leaf area, time series of weights of organs, waiting times for first occurrences. It is then hard to assume independent identical normal errors or something equally simple. A more refined model is required to somehow reflect the different types of data. Duplicate measurements are useful to allow formulation of a model for random effects.

The differentiability requirement is the least fundamental of the three conditions. Approximate normality of the estimator will be lost in case of non-differentiability. Moreover, many common tools for finding the estimator and assessing its accuracy require differentiability. But apart from that, non-differentiability can be overcome by computer power. In contrast, non-identifiability reflects a serious data shortage, in amount or in quality. This might be overcome, however, when data are supplemented with additional information. For all statistical approaches, a model for the randomness in the observations is essential.

It will be shown that in case of non-identifiability, the Bayesian approach allows distributional parameter estimation when observational data are supplemented with sufficient prior information on the parameters. The only additional requirement is a model for the randomness in the data. In theory, the Bayesian solution is simple and elegant, but the actual calculation of the estimator may pose formidable numerical problems.

We do not claim that the Bayesian method is the only solution to identifiability problems. In particular, a set theoretic approach may be attractive when for some reason it is not possible to formulate a plausible randomness model (Keesman and Van Straten 1990).

4.3 Bayes' rule

Bayes was a British statistician, living in the 18th century. He lived a rather quiet life and his scientific papers were only published after his death. His main contribution to science has been the development of the so-called Bayes' rule. Bayes' rule relates prior probabilities to posterior probabilities. We will now investigate how Bayes' rule can be applied to model calibration. A challenge is to find reliable estimates for the $\Pr(A_i)$, being usually based upon either subjective knowledge or knowledge from previous experiments. We will therefore return to the model equation $y = f(x,\theta) + \varepsilon$.

Denote the prior density of parameter vector θ by $p(\theta)$. This could be the homogeneous distribution on an interval. Hence, the prior probability that θ lies in a domain D of the parameter space is given by $\int_D p(\theta)\, d\theta$. New data y enter through the likelihood function, $l(\theta|y)$. Formal definitions of the likelihood are given in many statistics textbooks (see for example Cox and Hinkley 1974). The function $l(\theta|y)$ is calculated from the data y, model predictions at some parameter value θ, and a stochastic model for the differences between predictions and observations. In Section 4.6 we will show how to calculate the likelihood function in a common calibration situation. The prior information $p(\theta)$ and observations y are combined into a posterior density function $q(\theta|y)$ by Bayes' rule:

$$q(\theta|y) = c\, p(\theta)\, l(\theta|y).$$

Note that the posterior density depends on the observations y. The constant c is determined by the requirement that the integral of $q(\theta|y)$ over parameter space equals 1. Therefore, given the prior information on θ and the relations between θ and y we now have the density of θ given y. The posterior probability that θ lies in a domain D is given by $\int_D q(\theta|y)\, d\theta$. Informative data should lead to a posterior distribution more peaked than the prior, hence giving more precise information than available as a prior. When more observations become available, the posterior $q(\theta|y)$ can take the role of prior in an iterative process: Bayes' rule describes a step in a learning process.

The problem with Bayesian statistics is that one always needs a prior, in particular at the first application of the rule, when nothing is known yet. One starts therefore usually with an uninformative prior, e.g. the normal distribution with a large standard deviation. Also expert knowledge may be used in the choice of the first prior.

Bayes' rule provides an elegant, theoretically simple, answer to calibration. Before we come to that, however, we will first discuss how to construct priors that summarize the results from literature, and how to evaluate whether a prior distribution available from the literature prior suffices for predictions.

4.3.1 Constructing a prior from literature reports

By way of example, we consider the case of a single parameter θ that may differ between different situations. This natural variability of the parameter is modeled as randomness. Suppose the parameter has been measured or estimated in n situations that form a random sample from a population. Parameter values are denoted by $\theta_1,\dots,\theta_n$, and measurements by $y_1,\dots,y_n$, with reported standard deviations $s_1,\dots,s_n$. Such a list may be the result of a literature search. The case of a vector of parameters is more complicated, but can be treated similarly.

Parameters θ_i are independent with mean and standard deviation, say μ and τ, both unknown. They have been measured with error say ε_i, with different standard errors σ_i, which have been estimated by s_i. The accuracy, or number of degrees of freedom, of s_i is not known. In view of this lack of information, it will be assumed that s_i is perfectly accurate, i.e. $s_i=\sigma_i$. The main problem is to predict θ_{new}, the parameter value in a randomly chosen new situation. A related problem is to estimate θ_i in a situation that has been measured but where model validation requires its distribution. The problem may be solved in a classical and a Bayesian way. In the next example we shall briefly discuss a Bayesian solution, as described for instance in Gelman et al (1995).

4.3.2 Example

A model parameter has been measured in 9 situations. The measurements y_i and standard errors s_i are given in Table 4.1.

i	1	2	3	4	5	6	7	8	9
y_i	8.8	9.0	9.1	9.7	9.8	10.1	10.2	10.7	11.6
s_i	1.8	1.0	1.2	1.3	1.1	2.0	1.0	1.2	1.6

Table 4.1. Measurements y_i of parameter θ_i in nine situations with standard errors s_i (i=1...9).

The analysis reported below assumes that parameters θ_i, and measurement errors ε_i are normally distributed. More refined assumptions, for instance with specific distributions for parameters that have to be positive (which is not honored by assuming a normal distribution), can also be implemented; but we want to keep this example simple. Standard uninformative priors are used in the analysis (Gelman et al 1995).

Some quantiles of the posterior prediction of the parameter value (not its measurement) in a new situation, and in the ninth situation are tabulated below.

Cumulative probability	0.025	0.250	0.500	0.750	0.975
θ_{new}	8.26	9.45	9.89	10.26	11.65
θ_9	8.78	9.64	9.99	10.41	11.84

Table 4.2. Quantiles of posterior distributions of parameter θ in a new situation (θ_{new}), and in the 9th situation (θ_9) from the data set.

The variance of θ_{new} equals 0.6816, which is less than the variance, 0.7961, of the measurements. Note that the 95% interval of θ_9 is smaller than the interval that one may construct from information about the ninth situation only, namely $y_9 \pm 1.96\, s_9$.

If one can be certain that the studied parameter is constant in the whole population, that is τ^2=0, one should use that knowledge. The posterior distribution for the parameter has a variance equal to 0.1737: the knowledge is rewarded with a much sharper estimate.

4.4 Uncertainty analysis

On the one hand models are being developed, whereas on the other hand they are intended to be used to make predictions. Uncertainty analysis estimates prediction uncertainty caused by given uncertainty of model input: parameters, initial values and exogenous circumstances. Structural errors in the model are beyond the scope of uncertainty analysis. Uncertainty analysis aims to identify input variables, or groups of input variables, that contribute most to prediction uncertainty. The results of such an analysis can help to set research priorities. Often uncertainty analysis proceeds from a probabilistic description of input uncertainties. Bayesian parameter estimation forms an excellent starting point for such an uncertainty analysis.

Metselaar (1999) performed an uncertainty analysis for SUCROS-MAIZE, based on parameter uncertainty remaining after a thorough literature study. The resulting prediction uncertainty appeared to be too large for practical application of the model thus parameterized. Apparently, literature information falls short to parameterize the model, showing a need for calibration.

4.5 Bayesian calibration

4.5.1 Background

Suppose that analysis of literature data has lead to a distributional prediction of parameter values in a new situation. Often it will appear that this distribution is too vague for making precise predictions. But suppose that independent observations are available for that particular situation and a stochastic model for the error in those measurements. One may wish to combine the two forms of information to obtain more precise predictions. Obviously, there is no guarantee that the predictions will become sufficiently precise. Moreover, a precise prediction need not be a good prediction, since model and additional assumptions may be erroneous: this will be discussed in section 4.6.

The posterior parameter distribution derived from the literature will serve as a prior distribution in the calibration. The error model allows to calculate a likelihood function. Consider the formulation as described in Section 4.2, with observation vector y modeled as $f(x,\theta)+\varepsilon$, where ε has probability density $g(\varepsilon)$. Thus, the likelihood has the form $l(\theta|y) = g(y-f(x,\theta))$, and is a function of the residuals. Thus a model run is required for each calculation of l. Usually this run takes nearly all the time required to calculate l. The prior information, $p(\theta)$, i.e. the probability density of the parameter vector θ, and the new data, y, are combined into the posterior density function, $q(\theta|y)$ by Bayes' rule: $q(\theta|y) = c\, p(\theta)\, l(\theta|y)$, as described in Section 4.3. Discontinuities in the likelihood, caused by discontinuities in model predictions, cause no problems in the formulation of the rule. Therefore the 2nd requirement in section 4.2. (differentiability) is not required for a Bayesian calibration. Theoretically impossible parameter values can be excluded by zero prior values. Presence of prior information also prevents problems with identifiability. An uncertainty analysis with posterior probabilities allows to assess whether or not the posterior prediction uncertainty is acceptably small under specific new exogenous circumstances.

The remainder of this section is devoted to numerical implementation. Implementation will appear to be far from simple, in particular for large numbers of parameters.

4.5.2 Numerical implementation 1: the grid method

Reilly (1976) and Aldenberg et al (1995) implement Bayes' rule in a simple, intuitively appealing way by discretizing the parameter space. Each of the say k parameters is restricted to a discrete set of say m possible values, yielding a grid of m^k points. Subsequently, the values of prior and likelihood are calculated at each grid point. Calculation of the likelihood therefore involves m^k model runs. The grid method will become prohibitively computer-intensive with more than a few parameters.

4.5.3 Numerical implementation 2: Monte Carlo methods

Monte Carlo methods provide alternatives to the grid method. Their aim is to produce a sample from the posterior distribution. Two typical instances will be discussed.

Rejection sampling

Rejection sampling, the simplest Monte Carlo method, works as follows. Let L denote an upper bound of the likelihood function $l(\theta)$ for that part of the parameter space Θ where the prior density is non-zero. A parameter vector θ is drawn at random many times from the prior distribution and the likelihood $l(\theta)$ is calculated. The parameter vector is accepted with probability $l(\theta)/L$. The procedure is most efficient when L is the maximum of the likelihood. The accepted parameter vectors constitute an independent random sample from the posterior distribution. The method is feasible if the average acceptance probability is sufficiently large, but might become very computer-intensive with small acceptance probabilities.

Markov Chain Monte Carlo

The Metropolis method is the original form of Metropolis-Hastings type Markov Chain Monte Carlo methods (Gelman et al 1995).

The chain starts with an arbitrary parameter vector with non-zero prior probability and non-zero likelihood. Denote the latest parameter vector in the chain by θ, with prior $p(\theta)$ and likelihood $l(\theta)$. Then the next vector is obtained as follows. One draws a candidate point, say θ', by some random proposal generating mechanism: for instance $\theta'=\theta+\eta$, with η multivariate normal. The candidate θ' takes the next place in the chain by a random mechanism: it is accepted when

$$\frac{p(\theta')\cdot l(\theta')}{p(\theta)\cdot l(\theta)} > u$$

in which u is random with uniform distribution on the interval (0, 1). Thus, higher posterior densities are always accepted, whereas lower probabilities can be accepted as well, but with a lower probability. If θ' is rejected, θ will succeed itself. The chain thus constructed will converge to a stationary chain with elements drawn at random, but not independently, from the posterior distribution.

The performance of the Markov Chain Monte Carlo method depends critically on the proposal generating mechanism. The main problems of the procedure are the choice an efficient proposal-generating mechanism and monitoring the performance of the chain.

4.5.4 Comparison of grid method and MC methods

An attractive feature of the Monte Carlo methods is that inclusion of a parameter with little or no influence on model outputs will barely affect the acceptance probability. It will therefore hardly if ever lead to additional calculations. In an informal sense, the workload seems to be proportional to the information content of the new data. With the grid method, in contrast, each additional parameter multiplies the workload.

A disadvantage of the Monte Carlo methods is that it is hard to assess the required computational effort in advance, without actually performing some sampling. This effort might be evaluated in a sampling experiment of moderate size.

4.5.5 Example: an exercise with SUCROS

The acronym SUCROS stands for 'Simple and Universal CROp growth Simulation model' (Spitters et al 1989). From the statistical point of view, however, the model is still quite complex: it has about 50 parameters. SUCROS is a dynamic model in discrete time, namely days. Daily radiation and minimum and maximum temperatures are exogenous input. The major state variables are the development stage, the leaf area index and the dry weights of various organs.

In the course of time, the crop passes through consecutive phases and the dynamical laws change discontinuously from phase to phase. The parameter-response-functions have jump-discontinuities due to the distinct growing phases (Metselaar 1999). These cause complications during sensitivity analysis and calibration and therefore classical Taylor approximations and asymptotic statistical estimation theory do not apply.

The following example provides a clear case of non-identifiability. Eight parameters were calibrated with prior information from a literature search (Metselaar 1999). The, artificial, system-specific data consist merely of a single measurement of total above-ground dry-weight of maize at harvest in Wageningen in 1985: 15 ton per ha, the error following a normal distribution with standard deviation 1. This measurement can be fitted with many different parameter settings. Therefore, classical calibration methods for more than one parameter are deemed to fail. A Bayesian posterior sample of size 1000 was constructed with the Monte Carlo rejection method.

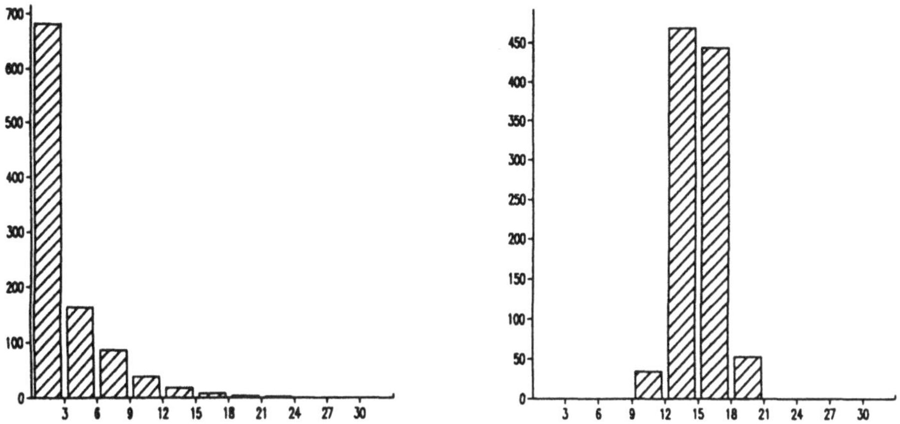

Figure 4.1 Histograms of samples of size 1000 of prior and posterior predictive distributions of total above-ground dry-weight (ton per ha) of maize in Wageningen in 1986; left: prior; right: posterior.

Figure 4.1 shows prior and posterior predictive samples of total above-ground dry weight at the same location under different weather circumstances, namely those of 1986. The uniform prior is useless for prediction; its low mean reflects that the prior comprises variability of maize cultivars over the world, whereas the cultivars actually used in The Netherlands have been selected for suitability to the local climate. The posterior prediction is much sharper: apparently the measurement from the previous year adds valuable information, which can be recovered by the Bayesian calibration procedure.

4.6 Model validation

Model validation is a multifaceted subject; see for instance the special issue of Advances in Water Resources (Hassanizadeh and Carrera 1992). This section will touch upon the difference between validation of a point prediction and a probabilistic validation based upon Bayesian calculus of parameter uncertainty.

We use the term validation in the utilitarian sense of characterizing prediction error of a calibrated model in a specified population of situations. With point predictions, the mean squared prediction error constitutes an obvious characterization of prediction error, the mean being taken over a random sample from the population. The mean squared prediction error incorporates input uncertainty, structural error, and error in the validation data measurements (Wallach and Goffinet 1989). But a problem remains how to express the quality of a probabilistic prediction. It is easy to make a probabilistic prediction so vague that it is never entirely wrong, in the sense that any outcome is predicted to be possible. Such vague predictions should get a low score. On the other hand, the score should discourage sharp predictions when the outcome is known to be very uncertain. The following measure for the quality of a probabilistic prediction is based upon these considerations. Let $\Pi(z)$ denote the cumulative distribution function of the prediction; i.e. Prob($z < z_0$) = $\Pi(z_0)$. Let y denote the outcome, and let $U_y(z)$ denote the cumulative distribution of the outcome, that is, $U_y(z) = 0$ when $z < y$ and 1 when $z \geq y$. Then the quality of the prediction may be expressed by the loss function

$$L(\Pi,y) = \int (U_y(z)-\Pi(z))^2 \, dz$$

(see Figure 4.2). The loss function is the sqaure of the differences between the distribution of the outcomes and of the predictions. Loss is small if $\Pi(z)$, the distribution of the predictions, is close to $U_y(z)$, the distribution of the outcomes, for all values of z, i.e. if the model outcomes are close to the predictions for the whole range of values of interest. Note that for a point prediction, the loss function is equal to the absolute value of the prediction error. Loss functions for probabilistic predictions are discussed in more detail in Kroese et al (1995).

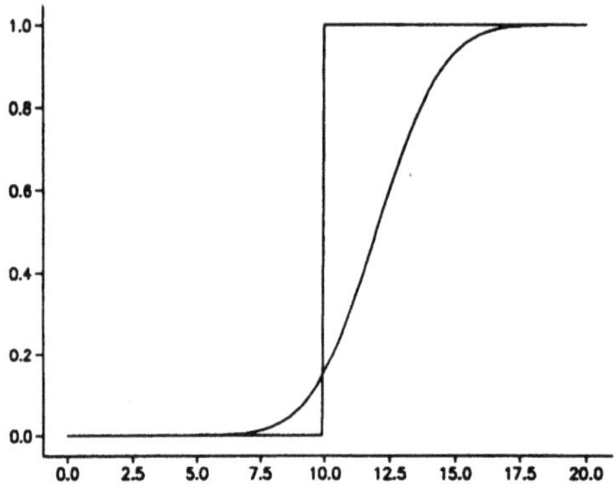

Figure 4.2. Loss function for probabilistic prediction. The smooth curve represents the cumulative distribution of the prediction. The step function depicts the cumulative distribution of the observation (which equals 10 in this graph). The loss is the integral of the squared difference between the two functions.

4.7 Decision support

Modeling is often used to support decisions. In theory, the model is used to predict the behavior of a system under various alternative actions. The decision maker chooses the action leading to the most desirable predicted behavior. Practice, however, is often more complicated: the decision maker may not have absolute trust in the model and its parameterization, the model rarely covers all relevant aspects, and it will often be hard to decide which outcome is most desirable. For such reasons, decisions are not *calculated*, but merely *supported*.

A Bayesian prediction takes the form of a probability distribution, usually multivariate, that represents uncertainty about the system's behavior. Thus, one shortcoming of deterministic predictions for use in decision support is avoided. Different actions lead to different distributions, so the decision maker has to choose between distributions. There is a rich literature on Bayesian decision theory; see for instance Berger (1985).

4.8 Discussion

Bayesian methods are suitable to update existing information about model parameters when new information becomes available. The information about parameters is represented by a probability distribution. Prediction uncertainty under specific exogenous circumstances can be evaluated by means of Monte Carlo simulation.

This chapter discussed the possibilities for assessment of post-calibration uncertainty of a parameter-rich model as an integrated part of the calibration. The calibration was based on a prior parameter distribution and new observations. It was seen that this calibration problem has a simple theoretical solution, when one has a stochastic model for the difference between prediction and observations. Prior information prevents identifiability difficulties. The actual computation might become prohibitively time-consuming with increasing number of parameters.

From a theoretical point of view, the computer-intensive nature of Bayesian calibration is a mere technical problem. More fundamental problems may well arise that throw doubt on the applicability of the method, for instance through errors in the model structure, in the prior information, or in the stochastic model for the residuals. The relevance of the prior is sometimes controversial: it may for instance be questioned whether processes studied in laboratory experiments really are the same as those occurring outside. When calibrating parameter-rich models, one often fixes some moderately sensitive parameters in advance. The ensuing error will cause difficulties in the formulation of a plausible stochastic model for the residuals. In cases where the basic assumptions are very doubtful, the calibration procedure is no more strongly directed by such assumptions, which may lead to some embarrassment of choice.

Bayesian calibration offers prospects to enhance the usefulness of model predictions through an indication of their precision. It might sometimes happen, however, that the basic assumptions of the calibration are doubtful, or that the procedure is too computer-intensive. In such cases, prediction quality may still be evaluated by means of independent validation data. Anyhow, the mere fact that the basic assumptions are numerous, constitutes an argument to validate on independent data.

Exercises

1. From the example 4.3.2. construct the likelihood functions on the basis of identically and normally distributed errors. Use, if necessary, the provided data. Sketch how estimates for θ_{new} and θ_9 can be obtained.

2. Consider one of the global radiation models in chapter 3. Discuss elements of identifiability, differentiability and presence of a randomness model. Are there any major differences between the three models?

3. Sketch how a Bayesian calibration can be carried out. Implement this procedure in a worksheet.

4. Create a formal chart in which support of decisions are clarified. Take as a decision the amount of nitrate to be applied. Indicate where models, prior probabilities and Bayes' rule can be applied.

5 Models and Scale: Up- and Down-Scaling

JU Smith

5.1 Introduction

The need to change the scale of a developed is a general problem in systems modeling. In particular, with the further development of modeling tools to solve problems posed by policy makers, advisers and farmers, modeling systems that can function at a number of different scales are increasingly needed. Such systems should be able to determine the optimum scale of operation, and even to select the most appropriate model to use. The aim of this chapter is to derive some generally applicable methods for up-and down-scaling from the area of modeling nitrogen and carbon behavior in soil/crop systems. This may allow a model to be adopted at the scale at which data are available, and results are needed.

In systems modeling, changing scale is often defined in terms of changes to the spatial area: up-scaling is an increase and down-scaling is a decrease in the size of the spatial area units. A small scale refers to large spatial units, whereas a large scale refers to small spatial units. We will use a more general definition. Changing the scale of a model is defined as a change to the resolution of *any* output values from the model. Changes in scale therefore may involve changing the resolution of any of the model input variables. We will discuss the underlying scientific and modeling developments required to maintain model integrity when changing the resolution of input and output values. Examples are

- changes in spatial units - crop type from crop·ha^{-1} to crop·5km^{-2};
- temporal units - rate of mineralization from kg·ha^{-1}·day^{-1} to kg·ha^{-1}·week^{-1};
- yield units - nitrogen requirement of the crop from (kg N)·(t Grain)$^{-1}$ to (g N)·(t Grain)$^{-1}$.

Leffelaar (1990) discussed the level of organization to which a problem should most appropriately be analyzed. He itemized criteria to consider when determining the scale at which to perform simulations. These include objectives of the study, level of organization of the scale or processes, scale of scientific understanding, possibility of clearly defining subsystems, possibility of validating the model of a system and possibility of returning to the field level of organization in the systems synthesis phase. Working within these criteria, two practical reasons for changing scale emerge: *transferability* of the model to novel environments and model *robustness*.

> **Two practical reasons for changing scale emerge: *transferability* of the model to novel environments and model *robustness***

An up-scaled model is often more reliably transferred to novel environments than a model developed with a low resolution. If a model describes the system using accurate scientific paradigms of the underlying processes, the effects on each process of changing the environment will be simulated, and so the accuracy of the model will be maintained. It is often difficult to determine the underlying mechanisms at the larger scale because there is no opportunity for evaluating the simulation of each process. Therefore it becomes necessary to establish a description of the process mechanisms within a smaller scale model and then adapt the model for use at the larger scale.

A down-scaled model, however, is often more robust than a model developed with a higher resolution. Our understanding of a system can be greatly enhanced by characterizing its behavior using statistical methods alone. General rules may be derived to apply at the larger scale. Although it may not be transferable to novel environments, such a model is correct within the limits of its development. The statistically derived rules may then be adapted, for instance using scientifically meaningful multiplication constants, so that the model can be used at the smaller scale. The scaled-down model maintains overall accuracy at the large scale, because it is not subject to propagation of errors from the individually modeled sub-processes. As a result, the robustness of the larger scale model should confer robustness to the scaled-down model.

In practice, however, the scale of system operation is dictated by the questions being asked of the system and the data available. As an example, a modeling system incorporating the dynamic field-based nitrogen-turnover model, SUNDIAL (Bradbury et al 1993; Smith et al 1996a) and the static catchment nitrogen model, NEAP-N (Miles et al 1995; Anthony et al 1996) is asked the questions "What is the peak nitrate concentration of nitrate leached from arable land?" and "When and where will the peak occur?". The questions demand that the models simulate peak nitrate concentration on a daily, weekly, or at least monthly basis for the field. If the attached database holds information about the dominant soil and land cover type on only a 1 × 1 km grid-square (Howard et al 1995), the available data requires either that the system should run using 1 × 1 km grid-squares or that the data should be interpolated. The SUNDIAL model simulates nitrogen turnover with a weekly time-step, but is developed at the field scale. To use SUNDIAL with the available data either the data must be interpolated to the field scale or the spatial units of the model must be scaled-up to a 1 × 1 km grid-square. The NEAP-N model, conversely, is well adapted to simulate the nitrogen status of the soil on a 1 × 1 km grid-square, but with an annual time-step. To obtain the required result using NEAP-N, the temporal units of the model must be scaled-down to provide results on at least a monthly basis. One way of achieving this is through an algorithm developed to express peak concentration in terms of model parameters. The scale of system operation then depends upon the relative errors introduced by interpolation of data, spatial up-scaling or temporal down-scaling. The best methodology to use can only be determined by quantifying the likely errors and assessing the need for a model that is transferable to a novel environment.

Changes in scale may result in changes in the scope of the model. For example, gravitational leaching of water down the soil profile is not an important process in the soil micro aggregate and so may be omitted from an aggregate model (Leffelaar 1988). Scaling up from the aggregate to the plot requires the additional process of profile leaching to be included. The consequences of omitting or incorporating additional processes must be examined and methods devised to ensure model validity is maintained.

Changes in scale also result in changes in the *heterogeneity* of the values of input variables. Up-scaling reduces accuracy and increases heterogeneity. The increased heterogeneity may be continuous or discontinuous. For example, the variogram for volumetric water content indicates an increase in heterogeneity over a range of 2 to 600 m (Addiscott and Tuck 1996). Accurate up-scaling from lysimeter to field requires some method of accommodating this increased spatial variability, either by interpolating the input data or by adapting the model. Statistical methods for interpolating continuous data are discussed elsewhere (Webster and Oliver 1990). Methods for interpolating discontinuous data, however, are not so clearly characterized. An example of the latter is the definition of crop type over the 5 km^2 data units of a catchment model. There cannot be a continuous variation between crop species, but some method of describing the different crops incorporated in the data unit must nevertheless be established.

Down-scaling, by contrast, increases the accuracy needed in the input variables and reduces heterogeneity. In order to down-scale, the heterogeneity of the larger scale input variables must be established and the influence of each variable on the large-scale model partitioned accordingly. Determining the influence of each variable and resolving the partitioning of results may require as much or more scientific understanding than would be needed for development from inception of the model with higher resolution. However, down-scaling can confer greater robustness to the model, and so both approaches are valid.

Data requirements are generally greater for a model developed to describe processes with high resolution than with low resolution. There is often insufficient data to run and evaluate a detailed model at a larger scale. Scaling-up demands either that high resolution data requirements are fulfilled or the model is adapted to reduce the data requirement (Smith 1995a). Methods of estimating input variables from the available data must be developed if the detailed model is to be applied unchanged. The accuracy of simulations achieved by simplifying the model or interpolating data should be compared when deciding on the best strategy for changing scale.

If a small scale model is used to simulate processes at a larger scale, the efficiency of the simulation must be maximized. Lord (pers.comm.) calculated that if a field-scale nitrogen model was applied in a national calculation on a field-by-field basis, the computer processor time required for the calculation would be prohibitively high. In addition, data entry would be extremely time-consuming. Methods must therefore be developed to reduce the number of unit calculations necessary and increase the efficiency of data entry.

In the following text, potential solutions to the problems associated with changes in scope, heterogeneity, data requirement and computing efficiency will be discussed in the context of ongoing work to develop a new system for UK nitrate policy support, required to function at scales ranging from field and farm through to catchment, regional and national level (Powlson and Mirza 1997).

5.2 Changes in scope

5.2.1 Determining Changes in Scope

A simple, but comprehensive and systematic procedure can be used for determining all changes in scope associated with changing scale (Fig.5.1). This procedure is centered around drawing up lists to describe characteristics of the system. The first stage is to list all questions that are driving the change in scale. This allows the output values required to answer each question to be inferred. All processes included in the model are listed, and processes that provide required values are marked. Any required output value that is provided by none of the processes represents a necessary increase in the scope of the model. The model is then adapted to incorporate the additional processes.

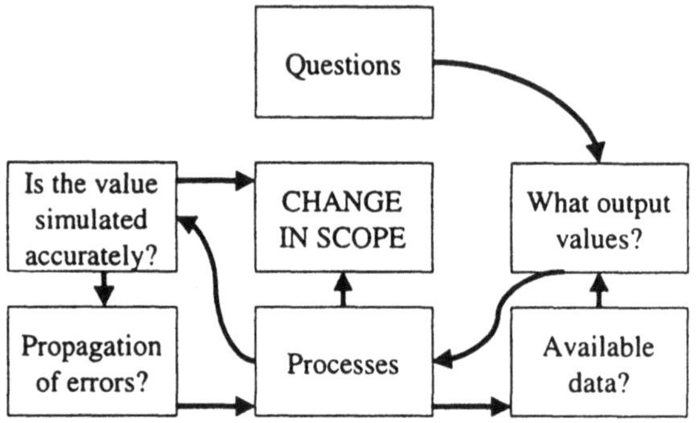

Fig. 5.1 Changes in scope associated with changing scale

The data available to drive the simulation are listed and the input values required by each marked process itemized. Inputs not included in the list of available data are added to the list output values that must be produced by some part of the model. Again, the processes that output each required value are marked, and the model is adapted to incorporate any further increases in scope necessary.

This procedure continues until all required input values are included in the list of available data. Any process that influences none of the required output values represents a decrease in the scope, and the model is adapted to omit the redundant process.

The next stage in determining the change in scope is to evaluate the accuracy of each identified output value simulated by the new model. Failure of the simulation may be due to error propagation through the model from the newly incorporated processes (Leenhardt 1995). The error ($Err_{F(a)}$) from the newly incorporated process ($F(a)$) must be lower than the acceptable error (Err_{acc}) less the error ($Err_{G(F(a))}$) arising from the action of the rest of the model ($G(F(a))$) i.e.

$$G\left(F(a)+Err_{F(a)}\right) < G\left(F(a)\right) + Err_{acc} - Err_{G(F(a))}$$

Methods for determining the acceptable error are discussed by Smith et al (1996b,c). If the standard errors of measurements are given, acceptable error may be defined statistically. Otherwise, the definition relies on arbitrary criteria.

Accuracy required in simulation of incorporated processes to achieve an overall result within the acceptable error may be determined by a simple sensitivity analysis in which the process module is substituted by an input variable, I. The sensitivity analysis allows the acceptable range of input values, $(I + Err_I)$, to be obtained from the acceptable error as before, i.e.

$$G(I + Err_I) < G(I) + Err_{acc} - Err_{G(I)}$$

A more comprehensive method of estimating the uncertainty contributions from several independent multivariate sources is discussed by Jansen et al (1994). The error associated with the incorporated process must be within the acceptable range determined by the sensitivity analysis, i.e.

$$Err_{F(a)} < Err_I$$

If the errors in the incorporated processes are within the acceptable error, the failure of the simulation must be due to errors in the scientific concepts of the processes involved, and the model should be modified.

Having completed the sensitivity analysis, the scope of the model is further refined by omitting any processes that the sensitivity analysis suggests are redundant at the larger scale.

The scope of the model is further refined by omitting any processes that the sensitivity analysis suggests are redundant at the larger scale

5.2.2 Incorporating and Omitting Processes

The new system for UK nitrate policy support (Powlson and Mirza 1997) is designed to allow additional modules of processes to be easily incorporated as science develops. This requires that instead of describing a whole ecosystem with a complete model each process is characterized as far as possible by an independent module. The ecosystem is then simulated by combining appropriate modules on a communicating framework.

Ensuring that each process is characterized by an independent module requires individual evaluation of process modules. A complete model can often give a valid result because errors in the underlying processes compensate. This is especially likely to occur when a model has been constructed using optimization procedures to obtain parameters that cannot easily be measured (Addiscott et al 1996; Smith 1995b).

Ensuring that each process is characterized by an independent module requires individual evaluation of process modules

Effectively, a portion of the description of one process is subsumed into the description of another. If the process module is omitted from the model, errors are introduced by the remaining portion. It may be difficult to obtain data to evaluate the individual processes, and so a model is often not comprehensively evaluated in all of its components. It is important, in this case, that the model is not further sub-divided than evaluation allows.

The communicating framework processes the questions entered into the system and calls the module that will provide the most accurate answer given available data and time for the calculation. The framework will request the information demanded by the module from the most appropriate source: an attached database or another module. Mounting sub-divided models on such a framework, allows the strengths of one model to be borrowed by another. Work is ongoing to establish the precise nature of links between modules in the new system for UK nitrate policy support (Cropper et al 1996).

5.2.3 Model Evaluation

Each of the 10 quantitative methods described in this section provides information on a distinct aspect of the accuracy of the simulation. The method used to assess the goodness-of-fit between simulated and measured values depends on the type of measurements available (Fig. 5.2).

In the equations below the following symbols are used.

Symbol	Explanation
O_i	observations
P_i	predictions (simulations)
$\overline{O}$	mean of the observations
$\overline{P}$	mean of the predictions (simulations)
n	number of paired values
N	number of experiments
m_j	number of replicates within each experiment
$\overline{O}_j$	mean of the measurements in the jth experiment
$\overline{d}_j$	mean deviation = $\overline{O}_j - P_j$

1. The total difference between the simulated and the measured values may be calculated as the root mean square error, *RMSE*, i.e.:

$$RMSE = \frac{100}{\overline{O}} \sqrt{\frac{1}{n} \sum_{i=1}^{n} (P_i - O_i)^2}$$

2. If standard errors of the measurements ($s_e(i)$) are available, the statistical significance of *RMSE* may be assessed by comparing to the value obtained assuming a deviation corresponding to the 95% confidence interval of the measurements (corrected from Smith et al 1996b), i.e.

$$RMSE_{0.95} = \frac{100}{\overline{O}} \sqrt{\frac{1}{n} \sum_{i=1}^{n} (t_{(n-2)0.95} \cdot s_e(i))^2}$$

where $t_{(n-2)0.95}$ is the Student's-t distribution for n-2 degrees of freedom with a two-tailed P-value of 0.05. An $RMSE$ value less than $RMSE_{0.95}$ indicates that the simulated values fall within the 95% confidence interval of the measurements.

3. If the experiments have been replicated, the lack of fit statistic, *LOFIT* (Whitmore 1991), should be calculated.

$$LOFIT = \sum_{j=1}^{N} m_j \bar{d}_j^2 = \sum_{j=1}^{N} m_j (\bar{O}_j - P_j)^2$$

The LOFIT statistic allows the experimental errors to be distinguished from the failure of the model.

4. If individual replicate measurements are available, the significance of *LOFIT* may be determined using an F-test, i.e.:
where *MSE* is the mean squared error. According to statistical convention, a

$$F = \frac{MSLOFIT}{MSE} = \frac{\sum_{j=1}^{N}(m_j - 1) \sum_{j=1}^{N} m_j(\bar{O}_j - P_j)^2}{N \sum_{j=1}^{N} \sum_{i=1}^{m_j} \left((O_{ij} - P_j)(\bar{O}_j - P_j)\right)^2}$$

value of F greater than the critical 5% F value may be taken to indicate that the total error in the simulated values is significantly greater than the error inherent in the measured values.

5. If no estimates of the standard errors are given, the accuracy of the simulation may be further assessed by calculating the modeling efficiency, *EF*, and the coefficient of determination, *CD* (Loague and Green 1991). The modeling efficiency, *EF*, provides a comparison of the efficiency of the chosen model to the efficiency of describing the data as the mean of the observations, i.e.:

$$EF = \frac{\left(\sum_{i=1}^{n}(O_i - \bar{O})^2 - \sum_{i=1}^{n}(P_i - O_i)^2\right)}{\sum_{i=1}^{n}(O_i - \bar{O})^2}$$

Values for *EF* can be positive or negative. A positive value indicates that the simulated values describe the trend in the measured data as effectively as the mean of the observations. A negative value indicates that the simulated values described the data less well than a mean of the observations.

6. The coefficient of determination, *CD*, is a measure of the proportion of the total variance in the observed data that is explained by the predicted data.

$$CD = \frac{\sum_{i=1}^{n}(O_i - \bar{O})^2}{\sum_{i=1}^{n}(P_i - \bar{O})^2}$$

The lowest value of *CD* is 0. A value of 1 or above indicates that the deviation of the predictions from the mean of the measured values is less than that observed in the measurements, i.e. the model describes the measured data as well as the mean of the measurements. A *CD* value less than 1 indicates that the deviation of the predictions from the mean of the measured values is greater than that observed in the measurements, i.e. the mean of the measurements describes the data better than the model. Taken together, *EF* and *CD* allow *RMSE* to be further interpreted where standard error values of the measurements are unavailable.

7. The bias in the total difference between simulations and measurements may be determined by calculating the relative error, *E* (Addiscott and Whitmore 1987):

$$E = \frac{100}{n}\sum_{i=1}^{n}\frac{O_i - P_i}{O_i}$$

8. If the standard errors of measurements are available, the significance of *E* may again be determined by comparing to the value obtained assuming a deviation corresponding to the 95% confidence interval of the measurements (corrected from Smith et al 1996b):

$$E_{0.95} = \frac{100}{n}\sum_{i=1}^{n}\frac{t_{(n-2)0.95}\cdot S_e(i)}{O_i}$$

An *E* value exceeding $E_{0.95}$ indicates that the bias in the simulation is greater than the 95% confidence interval of the measurements.

9. The nature of the bias may be further examined using the mean difference, *M* (Addiscott and Whitmore 1987):

$$M = \sum_{i=1}^{n}\frac{1}{n}(O_i - P_i)$$

The mean difference between measured and simulated values gives an indication of the bias in the simulation but is less informative than *E* because errors are not proportioned to the size of measurement. However, it is a useful statistic when standard error values are not available to derive a value for $E_{95\%}$ since *M* can be related directly to *t*:

$$t = \frac{M}{s_d/\sqrt{n}} = \frac{\sum_{i=1}^{n}(O_i - P_i)/n}{s_d/\sqrt{n}}$$

where s_d is the variance:

$$s_d^2 = \frac{\sum_{i=1}^{n}(d_i - M)^2}{n-1} = \frac{\sum_{i=1}^{n}((O_i - P_i) - (\sum_{i=1}^{n}\frac{1}{n}(O_i - P_i)))^2}{n-1}$$

The t statistic (here after Chatfield 1983) is used to show a significant difference between simulated and measured values. In line with statistical convention, a t value greater than the critical two-tailed 2.5% t value may be taken to indicate that the simulation shows a significant bias toward over- or under-estimation when compared to measured values.

10. To assess whether simulated values follow the same pattern as measured values, the sample correlation coefficient, r, may be calculated, i.e.:

$$r = \frac{\sum_{i=1}^{n}(O_i - \bar{O})(P_i - \bar{P})}{\left(\sum_{i=1}^{n}(O_i - \bar{O})^2\right)^{1/2}\left(\sum_{i=1}^{n}(P_i - \bar{P})^2\right)^{1/2}}$$

This statistic can be useful in assessing how well the shape of the simulation matches the shape of the measured data. However, if there is no clear trend in the measured data to give a spread of paired measured and simulated data values, the correlation coefficient is of only limited use in determining how well a model simulates the measured data.

The statistical procedures outlined in this section are summarized as a flow diagram in Figure 5.2.

5.3 Heterogeneity

5.3.1 Continuous heterogeneity

Continuously heterogeneous variables are often most conveniently scaled-up by considering the integrated sub-processes to be characterized by the mean (Smith 1995a). However, this assumption is only valid if the model response to the mean value of the inputs is the same as the mean of the model responses to all inputs individually, i.e.

$$\sum_{N} \frac{F(a_N)}{N} \equiv F\left(\frac{\sum_{N} a_N}{N}\right)$$

where $F(\ )$ represents the model response, a_N is the Nth input variable and N includes the whole population.

This is unlikely to be true, especially if the model is non-linear with respect to its parameters (Rao et al 1977; Addiscott 1993). Methods of obtaining a sample distribution of input values include the "bacon-slicer" (Addiscott and Wagenet 1985),

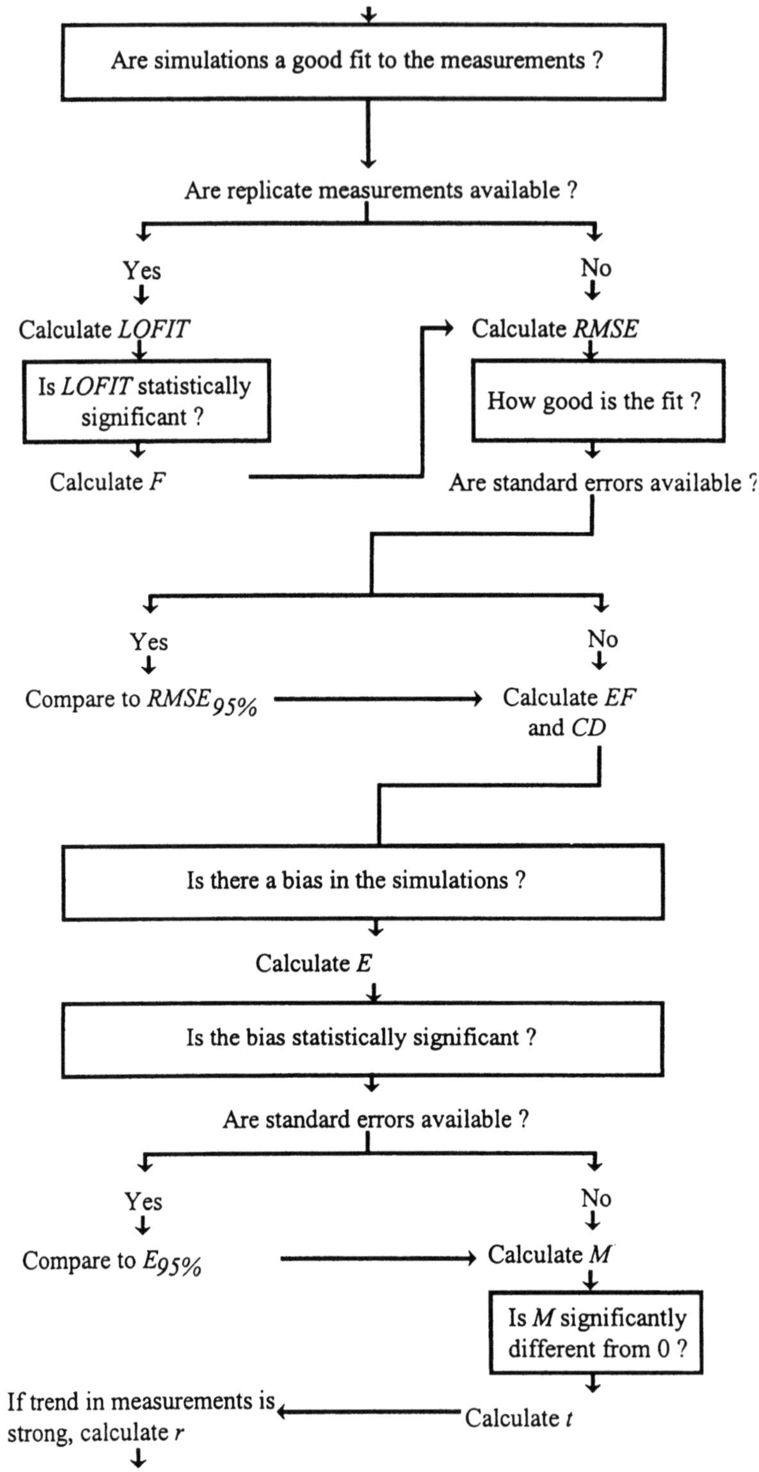

Fg. 5.2 Flow diagram for model checking and sensitivity analysis.

the Monte Carlo (Warrick et al 1977) and the Taylor series methods (Rao et al 1977). In this context, the "bacon-slicer" method is used for normally distributed populations because it requires only the mean and the standard deviation as input values and maintains the minimum number of necessary calculations. The normally distributed population is divided into slices with equal areas, and the mid-point of each slice is taken as the model input. Running the model using this sample population, provides a sample distribution of output values, equivalent to the distribution of the whole population. The mean of model responses to all inputs is then approximately equivalent to the mean of the sample output values, i.e.

$$\sum_{N} \frac{F(a_N)}{N} \approx \sum_{n} \frac{F(a_n)}{n}$$

and

$$F\left(\frac{\sum_N a_N}{N}\right) \approx F\left(\frac{\sum_n a_n}{n}\right)$$

where n includes only the sample population.

The assumption that the integrated sub-processes can be characterized by the mean is treated as valid only if the difference between the model response to the mean and the mean of model responses is less than the systematically defined acceptable error, Err_{acc} (Smith et al 1996b). The range of input values within which the acceptable error will not be exceeded can be calculated from the mean, the standard deviation and the acceptable error, i.e.

$$\sum_{n} \frac{F(a_n)}{n} < F\left(\frac{\sum_n a_n}{n}\right) + |Err_{acc}|$$

This then allows the applicability of the mean input variable to be assessed, simply by inspecting the standard deviation associated with each input value, and comparing to the acceptable range calculated for each input value, held in look-up tables within the modeling system. The value of other inputs may, of course, affect the acceptable range, and a sensitivity analysis should be done to determine how the value of other inputs influence the acceptable range. If the observed range of inputs is outside the acceptable range, the input cannot be characterized by the mean, and the simulation must be run for the sample distribution of inputs.

This automated up-scaling procedure may be further extended by measuring the variance of the input values as the scale increases (Stein et al 1989; Warrick et al 1990). By obtaining the relationship between the variance and the size of sampling unit, the limits for reliable up-scaling may be determined. Addiscott and Tuck (1996) discuss the influence of possible shapes of such variograms on the up-scaled model. Relationships describing the variograms may then be incorporated into the modeling system to allow the system to select the most appropriate model for the required accuracy, scale and availability of data.

5.3.2 Discontinuous heterogeneity

Discontinuously heterogeneous variables are most conveniently scaled-up by describing them as the majority value. For instance, crop type in a 1 × 1 km grid-square might be described as the majority crop type. This assumption is only valid if the model response to the majority value is statistically equivalent to the proportioned sum of the model responses to all possible inputs, i.e.:

$$\left| \sum_n p_{a_n} F(a_n) - F(a_{\max}) \right| < \left| Err_{acc} \right|$$

where p_{a_n} is the proportion of the population characterized as the discontinuous variable value a_n and a_{max} is the majority value.

Discontinuously heterogeneous variables are most conveniently scaled-up by describing them as the majority value

By calculating the contribution of each discontinuous input value to the overall result, the acceptable proportion of each may be determined. Again, the value of other inputs will influence the acceptable proportion of the discontinuous inputs, and a sensitivity analysis should be performed to determine how continuous input variables affect the acceptable proportion. The influence of other discontinuous input variables on the acceptable proportion may be determined by running the model for all possible combinations. The probability of obtaining each calculated output can be assigned by multiplying together the proportions of each input, i.e.

$$\left| \sum_m \sum_n p_{a_n} p_{b_m} F(a_n b_m) - \sum_m F(a_{\max} b_m) \right| < \left| Err_{acc} \right|$$

where p_{b_m} is the proportion of the population characterized as a second discontinuous variable value b_m.

This calculation assumes that the discontinuous variables are independent of each other. However, discontinuous variables are often related to each other. For instance, good management practices result in a regional relationship between crop type and soil type, a variable that is also often treated discontinuously. Such relationships can be incorporated and used to reduce the number of calculations needed. A system to determine all possible scenarios using management practices and current legislation as inputs is currently under development (Smith and Glendining 1996; Smith et al 1997a).

5.3.3 Partitioning the influence of heterogeneous input variables

Because down-scaling increases the accuracy needed in the input variables and reduces heterogeneity, the heterogeneity of the larger scale input variables must be established and the influence of each variable on the large-scale model partitioned. Rodda et al (1995) have down-scaled nitrate leaching calculated by the NCYCLE grassland model (Scholefield et al 1991) from annual to monthly values. In this approach they establish a relationship to weather patterns for plant uptake of N, N mineralization and gaseous loss of N through denitrification according to a soil classification based on maximum soil moisture deficit. The fraction of the resulting leachable N left in the soil is related to the drainage volume and the peak and average nitrate concentrations are related to the total N leached according to soil type and drainage status. Finally, using the assumption that peak nitrate concentration is associated with the first 100 mm of drainage, the timing of nitrate peaks is estimated to the nearest month by a simple soil water balance.

Smith et al (1998) have used an alternative method in which unknown input values to a dynamic model are obtained from an annual model that is driven by more readily available input values. Annual crop nitrogen requirement, expected yield, and return of organic manure calculated by NCYCLE, are used as inputs to the dynamic arable model, SUNDIAL (Bradbury et al 1993; Smith et al 1996a). The standard SUNDIAL algorithms and fixed parameters specific to a grassland crop are then used to calculate the weekly turnover of nitrogen under the grassland crop. This effectively uses the SUNDIAL model to partition the robust NCYCLE values. The robustness of NCYCLE is conferred on SUNDIAL by down-scaling, whilst maintaining the transferability of the process-based SUNDIAL simulation.

5.4 Data requirement

5.4.1 Adaptation of models to reduce data requirements

Many different methods are used to adapt a model to reduce the data requirements at the larger scale. The suitability of the different methods is dependent on the resources available and the precision of simulation required.

A very simple approach reduces the data requirements of a model by attaching a data-base containing national, regional and sub-regional means (Cropper et al 1996; Smith 1995a). The data requirements of the SUNDIAL model have been effectively reduced by attaching a database containing the sowing date, harvest date, expected yield, fertilizer application rate and fertilizer application date. If no field specific input values are provided, the modeling system automatically selects the values from the sub-regional, regional, or national database, depending on the detail with which the geographical location is specified. The reliability of using data at each scale may be assessed against look-up tables of the acceptable data range as described in section 5.3.

If the look-up tables indicate that mean values will provide an unreliable simulation, the model can be run using the mean value and mean plus or minus the standard deviation of each input, or using the bacon-slicer method described earlier (Addiscott and Wagenet 1985). These methods are more computationally expensive, but allow the error of the simulation to be quoted. The error increases as the scale increases, but the acceptable error also tends to increase with scale.

In an alternative method, the absence of direct measurements is overcome by calculating the input variable from some measured surrogate. Soil water properties are often estimated from easy-to-determine soil characteristics using pedotransfer functions (Salter et al 1966; Gupta and Larson 1979; Rawls et al 1982; Pucket et al 1985). Similar functions could be developed for estimating crop and organic manure properties (*phyto*transfer and *scato*transfer functions respectively?).

Anthony et al (1996) have developed a simple minimum information requirement (MIR) model for application at the catchment scale. The MIR is a simplification of the results of more complex models (e.g. SLIM: Addiscott and Whitmore 1991), attempting to retain the physical basis of simulations whilst reducing the dependence on input variables. The simplicity and comparative speed of the simulation should encourage a better understanding of the sensitivity of results to uncertainty in the inputs.

A simple minimum information requirement (MIR) model retains the physical basis of simulations whilst reducing the dependence on input variables.

5.4.2 Estimation of input data

A wide range of methods have been used to interpolate the input data necessary to run models. Jacucci et al (1993) developed a simple method for estimating soil characteristics from larger scale input values. Areal interpolation of land cover data for England and Wales, published annually at parish level, is described by Moxey (Moxey et al 1995). In this method the variation in land characteristics across parishes is used to model variation in land cover, thereby allowing the spatial basis of the parish data to be transformed and improving the description of distribution of land cover. Weather data may be interpolated from regional meteorological offices by a method developed by Semenov and Barrow (1996), which uses spatial regression down-scaling together with a local stochastic weather generator.

Weather data may be interpolated from regional meteorological offices by spatial regression down-scaling together with a local stochastic weather generator.

5.5 Computational logistics

If a modeling system is to be of any use to policy makers in decision support, the system should require a minimum amount of time for data loading and completion of simulations. The time required to run the system should be further reduced by attaching standard databases and using screen painting facilities to enter large bodies of data. The computational load represented by any model should be minimized by explicit design and efficient organization of the operations performed by in the modeling system. Methods to maximize computation efficiency are currently being incorporated into the new system for nitrate policy support (Powlson and Mirza 1997).

5.6 Conclusions

In the above text, guidelines for choosing and changing the scale of a model have been described. When changing scale, the developer should itemize and demonstrate any costs and benefits, e.g. reduced accuracy, increased transferability, increased robustness. The scale of the model that is chosen should be driven by data availability and the questions asked. The general approach described above uses the following protocol:

On up-scaling:

1. Decide which input variables will be scaled up. This usually includes all parameters with one type of denominator. For instance, all variables changing with time or all variables changing over space.
2. Determine any changes in scope.
3. Use a systematic method to define the required accuracy of simulations.
4. Decide if each variable changes continuously or discontinuously.
5. Perform a sensitivity analysis to assess whether difference between the model response to the mean of continuous input variables and the mean of the model responses is within the required accuracy: if it is, the mean input variable can be used in the simulation; otherwise, the distribution of the input variables must be entered.
6. Deal with discontinuous heterogeneity.
7. Adapt the model wherever possible to reduce the data requirement.
8. Where not possible, develop methods of filling in the holes.
9. Organize the computational system to reduce the computational load and to assist data entry.

On down-scaling:

1. Decide which variables will be scaled down. This usually includes all parameters with one type of denominator. For instance, all variables changing with time or all variables changing over space.
2. Perform control analysis to determine changes in scope.
3. Use a systematic method to define the required accuracy of simulations.
4. Decide if each variable changes continuously or discontinuously.
5. Perform control analysis to decide which variables influence variable to be scaled-down.
6. Partition the variables either according to determined relationships or by linking to a model that functions at a lower scale.

In practice, however, data may be limiting and so less rigorous methods of changing scale will be adopted. In addition, most models are developed as complete packages that cannot easily be divided and validated as sub-processes. As a consequence, many modelers are committed to working with the model already developed, and use such methods merely as a means of justifying the application of the model at another scale. By building flexible systems, based on modules of sub-processes, into which new modules can easily be added, it is to be hoped that this understandable personal bias will eventually disappear. Improved solutions to problems will be more quickly achieved using new modules which slot into such modeling systems, rather than by developing yet another complete model of the whole eco-system.

Exercises

1. Consider the following results of sets of experiments.

1st experiment			2nd experiment			3rd experiment		
I	O_i	P_i	*i*	O_i	P_i	*i*	O_i	P_i
1	23	22.5	1	27	23.4	1	23	22.4
2	25	22.9	2	21	22.1	2	24	22.7
3	24	22.7	3	20	21.8	3	25	23.0
4	22	22.3	4	23	22.4	4	23	22.3
5	23	22.4	5	25	22.8	5	26	23.2
6	23	22.5				6	22	22.2
						7	21	22.0

Determine for this experiment each of the error measures described in section 5.2. Values for the *t*-test can be derived from the attached table.

2. Consider the model on global radiation. Discuss the properties mentioned in section 5.3. Indicate clearly where testing is necessary and how it could be done.

Fig. 5.1 is a very general picture and needs to be evaluated for each particular model and each particular decision situation. Discuss how this can be done for the SUNDIAL model. Take care of data availability, models and the decision that has to be supported.

3. Changes in scope

A new module is introduced to an existing model. The new module (F()) outputs approximately the square of the input (a):

$$F(a) \cong a^2$$

The error associated with the calculation ($Err_{F(a)}$) is consistently 0.2 units. The output from the new module is passed as an input variable to the existing model (G()). The existing model outputs approximately a third of the input F(a)) to the power 3:

$$G(F(a))\,) \cong (F(a) / 3)^3$$

The error associated with the calculation ($Err_{G(F(a))}$) is consistently 0.3 units. The acceptable error for the whole calculation (Err_{acc}) is 0.5 units.
Test if the new module is sufficiently accurate given

1.1) a = 1.0 units
1.2) a = 1.5 units
1.3) a = 2.0 units

4. Model Evaluation

An long term agronomic experiment provides measured values (O_i) dating from 1970. A dynamic model is used to simulate the values (P_i) obtained in the experiment. The results are as follows:

Date	O_i	P_i	Date	O_i	P_i
1970	22.09	24.71	1980	20.58	25.34
1971	23.14	25.19	1981	21.60	25.75
1972	23.00	22.36	1983	21.60	26.20
1973	23.87	23.38	1985	22.23	24.55
1974	23.77	23.92	1987	21.77	23.69
1975	23.14	24.27	1989	21.07	24.23
1976	24.29	23.76	1991	20.20	23.83
1977	23.73	24.04	1993	23.87	24.07
1978	23.00	24.36			
1979	22.96	24.75			

4.1 Use the available experimental data to evaluate the model performance.

Searching through the archives uncovers the standard errors (SE_i) associated with each measurement as follows:

Date	O_i	SE_i	Date	O_i	SE_i
1970	22.09	0.67	1980	20.58	0.62
1971	23.14	0.56	1981	21.60	0.56
1972	23.00	0.62	1983	21.60	0.59
1973	23.87	0.40	1985	22.23	0.49
1974	23.77	0.73	1987	21.77	0.50
1975	23.14	0.56	1989	21.07	0.52
1976	24.29	0.69	1991	20.20	0.84
1977	23.73	0.64	1993	23.87	0.77
1978	23.00	0.49			
1979	22.96	0.50			

4.2 Use the standard errors of the measurements to improve the model evaluation.

A further search through the archives uncovers replicates O_{ij} of each measurement O_i as follows:

Date	O_i	O_{ij}					
		$j = 1$	$j = 2$	$j = 3$	$j = 4$	$j = 5$	$j = 6$
1970	22.09	21.84	23.10	24.36	22.68	20.16	20.37
1971	23.14	23.94	24.15	23.52	24.36	21.00	21.84
1972	23.00	23.73	23.52	24.57	23.94	20.79	21.42
1973	23.87	24.15	24.57	23.31	24.99	22.26	23.94
1974	23.77	24.36	26.67	24.15	23.52	21.63	22.26
1975	23.14	24.36	24.15	23.52	23.94	21.21	21.63
1976	24.29	25.62	25.83	24.99	24.15	23.94	21.21
1977	23.73	25.62	23.73	23.94	25.20	22.26	21.63
1978	23.00	24.57	23.10	23.52	23.31	22.47	21.00
1979	22.96	22.68	24.99	23.31	23.10	22.47	21.21
1980	20.58	20.79	22.47	21.84	20.58	18.27	19.53
1981	21.60	21.21	23.31	21.42	23.10	20.79	19.74
1983	21.60	22.26	24.15	21.00	21.21	21.00	19.95
1985	22.23	22.47	23.52	22.68	23.10	21.21	20.37
1987	21.77	22.68	23.10	21.84	21.63	21.84	19.53
1989	21.07	21.63	22.47	22.26	20.58	20.37	19.11
1991	20.20	20.58	23.56	21.00	19.95	18.06	18.06
1993	23.87	23.73	27.09	24.36	23.94	22.47	21.63

2.3) Use the replicates of the measurements to improve the model evaluation.

5. Heterogeneity

The following crop types, resulting in the given model output are observed in a region of land:

Crop Type (a_n)	Percentage of Total (p_{an})	Model Output ($F(a_n)$)
a	10	10
b	10	12
c	15	17
d	15	21
e	50	15

Given an acceptable error in the simulations (Err_{acc}) of 0.5 units, assess whether the crop type can be characterised as the majority crop.

III Geographical Information Systems for Production Ecology

Geographic Information Systems (GIS) are common tools to manipulate several layers of spatial information. Such layers could be of a large variety, and examples include soil properties, weather properties, hydrological properties and topographic information. But other spatial properties can be considered as well.

There are many reasons to consider the role of GIS for modelling. Many of the variables to be applied in models have a spatial extent, and show spatial variability to a larger of smaller degree. Hence, this information may be available in a GIS. Moreover, the data may vary within the study region, leading to different model evaluations at different locations. Further, models are usually in particular valid at those locations where measurements were taken. As these locations may differ between the variables, and also model evaluations are relevant at locations where no data were collected, there exists an obvious need to interpolate data towards areas of land. This interpolation is usually done by means of delineating areas in which the collected data can be considered to be uniform: individual fields or regions for soil properties, individual catchments for hydrological properties and well-defined sub-areas for meteorological properties, usually depending upon topography.

6 Modern geographic information systems and model linking

P A Longley and M F Goodchild

6.1 Introduction

Geographical analysis involves the use of a wide range of techniques and models to apply formal, usually quantitative structures to systems in which the prime variables of interest vary significantly across space (Longley and Batty 1996). Such techniques of spatial analysis form a well-defined subset of the larger set of analytic methods, defined by an invariance property: the results of spatial analysis are not invariant under changes in the locations of the objects being analyzed. The tradition has its historical roots in locational analysis (e.g. Haggett 1965), geostatistics and the theory of regionalized variables (Matheron 1971; Cressie 1991) and spatial statistics (e.g. Cliff and Ord 1973), yet there are many reasons why modern geographic information systems (GIS) are causing the field to undergo profound change. Today's GIS (broadly defined) provides a seamless computing environment for every stage of spatial analysis. In this chapter we assess the implications of this state-of-affairs, and its possible implications for the future development of spatial analysis in a GIS environment.

GIS is a fast-developing and multi-faceted technology for producing simplified digital depictions, or models, of the real world. Our ability to apply this technology to create rich depictions and images of geographical reality has been spurred by the innovation of new data capture technologies and the proliferation of secondary digital data sources. Any GIS-based model of the real world requires input of information, and the foundations to GIS-based analysis are provided by an emergent digital data infrastructure, assembled from diverse sources and based upon different data structures (Goodchild 1998). Data models of spatial distributions provide a link between raw data taken from the real world to the spatial analytical models that are used to create higher order generalizations. Much of spatial analysis as conventionally understood involves building 'models of models', and as such the outcome can be no better than the data and data models upon which it is founded. In this chapter we will argue that, in practice, no spatial analysis system is founded on solid rock: rather, the practice of spatial analysis requires us to sink piles into the shifting sands of empirical reality until they are secure enough to support the structure that we wish to erect. The visual medium of today's GIS is slicker and more sophisticated than ever before, yet this medium can in practice obscure a more ambivalent message: that the foundations to analysis are insufficiently robust and sensitive to context. The objective of this chapter is to describe how and why model linkage, in physical and social science alike, requires understanding of the digital data that provide the infrastructural framework to analysis.

In a general sense, infrastructure may be defined as 'the system of services (public and private) which provide a systematic framework for human living arrangements' (Longley 1998). We think of traditional physical infrastructure as being designed with function and foresight in mind and as being created in a rational and orderly manner. It is also fixed to a unique location on the Earth's surface, and as such the immediate spatial context to its usage is obvious and apparent. Yet there are a number of important differences between physical infrastructure as commonly conceived and the new emergent digital infrastructure to GIS-based analysis. The large and complex data sets available today may pertain to unique locations, but they have usually been captured using a range of generic technologies, for a multitude of purposes. They may also be shared between users who have no particular geographical affinities or understanding of sensitivity to context (Goodchild 1997). As such the emergent digital infrastructure to spatial analysis may be less rational and ordered than is desirable for particular applications. Yet GIS also provides us with means of tailoring multi-purpose generic data to the context of particular applications. Later in this chapter we will illustrate this point through case studies in the diverse areas of remote sensing (see Curran et al 1998 for more general arguments) and geodemographics (Harris 1999). If the analytical elegance and theoretical transparency of the spatial analytic tradition is to be reconciled with our messy empirical world, the digital infrastructure to GIS must be designed with the same considerations in mind as its physical counterpart. That is to say, it should be designed with function and foresight in mind and be built on clearly recognizable (though not necessarily conventional) foundations. In many respects the new digital infrastructure to GIS is less rational and ordered than we would like!

> GIS provides us with means of tailoring multi-purpose generic data to the context of particular applications

6.2 Spatial analysis in context: past and present

We have described the traditional practice of spatial analysis elsewhere (Goodchild and Longley 1999), and this section borrows from our previous review and interpretation.

6.2.1 The linear project design

Generations of students have been introduced to geographical research through being encouraged to hone their general interest in 'finding out' about something into tightly prescribed and clearly defined hypotheses. The myriad detail of geographical reality is of consummate interest, yet only a tiny part of real world complexity is 'researchable' within the resource constraints of a student dissertation or project. And in practice, the same kinds of resource constraints apply to all surveys and investigations of the real world. In the idealized linear schema shown in Fig. 6.1, research begins with the formulation of hypotheses, and culminates in their formal testing in the spirit of classical statistical inference. In between lies a clear and strict sequence of procedures – choosing a data collection method (and designing a survey schedule, as appropriate), identifying a sample design, piloting, field collection of data (with verification and resampling), collation of results, analysis, and report-writing. This sequence is strictly linear and independent of external events. Hypothesis formulation precedes the

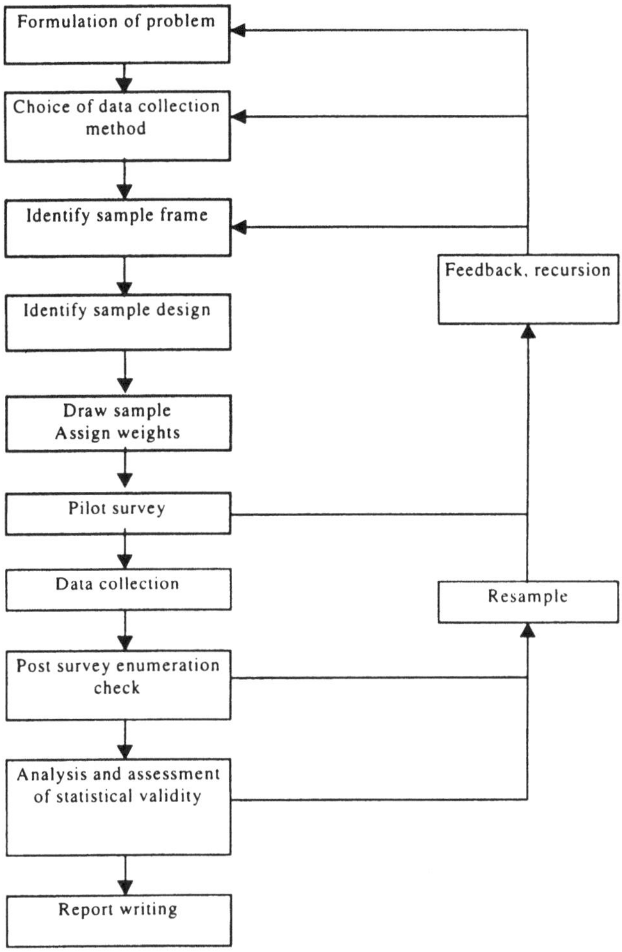

Figure 6.1. The traditional linear project design, with recursion. After Goodchild and Longley 1999.

collection of data and the performance of analysis. Availability of data has no influence on problem definition, and method of analysis is selected with the research hypotheses and characteristics of the new data set in mind (e.g. Levine 1981: Ch 17; Marascuilo and Levin 1983: inside cover; Siegel 1956: inside cover).

This 'linear project design' is to some extent recursive (e.g. major failures identified in the piloting phase may be fed into changes in the way in which the main data collection is carried out), and the search for hypothesis confirmation is reinforced by data exploration. Yet in no prevailing sense can it be described as data- or technique-**driven**—in stark contrast to 'data mining' approaches in which 'the data are allowed to speak for themselves', unconstrained by the conduits of hypothesis formulation. Traditional approaches to science have eschewed 'data-led thinking', instead taking a path towards understanding through cumulative, deductive reasoning.

6.2.2 Spatial analysis: retrospect

Implicit in our statement of the remit of spatial analysis at the start of this chapter is the view that the role of techniques and models is to help solve scientific or decision-making problems. The roots to the spatial analysis approach lie in the so-called 'Quantitative Revolution' in Anglo-American geography that was pioneered, *inter alia*, by Berry, Haggett, and Marble in the 1960s (e.g. Berry and Marble 1968; Haggett 1965).

The context to spatial analysis has been transformed over the last 30 years. The reasons for this are well known, and most obviously relate to the precipitous falls in the absolute cost of computer processing power. Looking back from the end of the twentieth century, the tools available to the first generation of quantitative geographers appear no more than rudimentary calculating devices capable of testing only the most restrictive of research hypotheses. The cost of computing in practice limited analysis to a single pass through the data in order to confirm (or refute) a research hypothesis. The rudimentary state of computing often required geographers to write their own (FORTRAN) programs (or adapt those of others), and thus researchers tended to have a specialist understanding of particular techniques rather than a general understanding of a wider range. And finally, the data that were generally available were restricted in quantity and quality by the budgets of particular projects and the lack of secondary georeferenced sources in digital form.

By the late 1980s, general technological changes in computing and the particular innovation of GIS had changed much of this. Increases in computing power led to routine exploration of data and multiple passes through data sets, while computer graphics made the spatial dimension to analysis much more explicit. The cumulative development of algorithms and subroutines led to the development of all-encompassing monolithic GIS packages with the widest range of analytical functionality. Together these developments encouraged spatial analysis in which the sensitivity of model forecasts could be measured against changes in control and predictor variables. These developments fundamentally separated the creation of spatial analytical functions from control of their usage.

As the 1990s progressed, two related processes characterized the subsequent development of GIS. First, the development of the Internet accelerated the break up of early GI systems into reusable (often application-specific) software modules which could be distributed across the Internet and be made available (through software vendors or other interested parties) as 'plug ins' to broader analytical functions. In some sense this development has brought spatial analysis back to the early years, in that responsibility for the design and precise configuration of particular applications has once again devolved to the analyst. It is not just software that has become distributed across the Internet, but also text (at individual and institutional Web-sites) and data: the context to spatial analysis has become digital. The second trend has been for software vendors to package 'fixed menu' offerings of analytical functions (and, increasingly, data) as low cost 'desktop GIS', permitting GIS-based models to be linked in a quite tightly prescribed range of ways.

These developments have each empowered spatial analysis but in rather different ways: the first has enabled specialist spatial analysts to assemble the widest range of subroutines and programs for everything from designing samples, through grossing of sample estimates, to analysis and report writing with text and graphics; while the second has introduced the novice to a guided and prespecified range of spatial analysis options which should be safe to use and tailored to specific research problems. Although different in conception and packaging, these different developments each represent an extension of the ranges of uses to which computers are put in spatial analysis.

From a historical perspective, the computing input to the linear research design of Fig. 6.1 initially involved limited batch processing of data. It developed into wider interactive data analysis with recursive feedback into hypothesis reformulation, and has today culminated in the use of computer software throughout every stage of research design and analysis—that is, everything from literature searches through to report writing and dissemination. Thus the computer has evolved from an equation-solving and data-handling device to a beginning-to-end environment for the modeling of complex systems and processes. Couclelis (1998) sees this not just as an incremental drift but rather as a sea change in which the IT environment in general has evolved from one in which comput**ing** entailed specialized numerical analysis at particular clearly-defined stages in research to one in which comput**ation** pervades almost all stages of research from hypothesis formulation to report-writing.

The computer has evolved from an equation-solving and data-handling device to a beginning-to-end environment for the modeling of complex systems and processes

6.2.3 Spatial analysis today: technique and application, induction and deduction

In spatial analysis, as elsewhere in science, those that have developed new techniques of analysis have historically devised appropriate means of testing and describing them. (Cliff and Ord 1973; Getis and Ord 1992) Yet even in the early days of computing, this world was changing, as the sharing of specialized subroutines between researchers and packaging of techniques into broader methods (informally, or through the development of subroutine libraries) obviated the need for every application to 'reinvent the methodological wheel'. Thus, for example, urban geographers seeking to classify urban social areas in the 1950s and 1960s shared principal components and factor analysis programs (and agreed upon appropriate distance and rotation methods), and it was not long before applications were spawned in almost every major city in the world.

In point of fact, this example illustrates three facets to the change that was taking place in computing. First, the techniques of classification became 'black box' tools which would generate outcomes ('results') for expert and non-expert users alike. Second, repeated use of an algorithm could gauge the sensitivity of classifications to particular combinations of input variables. Perhaps one of the more enduring legacies of the social area analysis approach, which otherwise amounted to a routine and unimaginative quest to identify communality of social structure between seemingly diverse case studies (see Berry 1971; Clark and Gleave 1973; Timms 1969), was that it suggested to human geographers that the measurement and choice of particular quantitative variables tightly prescribes the outcome of spatial analysis. And third, successive applications were associated with refinement of the basic clustering techniques, thereby leading to the cumulative development of computer algorithms and associated subroutines.

Together, then, cumulative developments in computing and technology led to increasing separation of technical development and application, and there was a redeployment of research effort away from development and programming of spatial analysis procedures for particular applications towards study of the sensitivity of packaged techniques to particular combinations of data and technique. In this way the deductive linear project design of Fig. 6.1 was gradually supplemented and to some extent replaced by inductive procedures designed to suggest empirical generalizations.

This repeated 'stimulus–response' approach to computing was not restricted to self-contained, stand-alone procedures such as principal components and factor analysis. The development of computer graphics was also associated with the development of procedures for exploring data (e.g. plots of residuals and potential leverage points), which strengthened the feedbacks shown in Fig. 6.1, and in turn led to the popularization of further new confirmatory analysis tools (such as resistant fit procedures). During the 1990s, a further wave of inductivism has run through spatial analysis with the development of WIMP (windows, icons, mice, pointers) computer interfaces, to the point at which today the medium of computation allows us to evaluate the messages of spatial analysis in (and through) the widest possible range of senses.

6.2.4 GIS and geocomputation

Although much of the preceding discussion has been framed in terms of the implications of general changes in the information technology environment for spatial modeling, they are clearly pertinent to any discussion of GIS. For example, the WIMP environment is particularly conducive to the conception and formulation of spatial problems, while today's network servers bring geographically enabled analysis to orders of magnitudes more users than ever before. As with mainstream computing, GIS today provides an open environment for every stage of any spatial analysis project. And the development of better user interfaces has led to the use of GIS in an ever wider range of applications contexts (Longley et al 1999 report estimates of the value of the digital geographical information industry between $10 billion and $14 billion by the mid 1990s). Popularization and diversification has also led to the blending of GIS with other background technologies: today many of the routine functions of spatial analysis are as accessible and easy to use as word-processing, particularly through standard software such as geographically-enabled spreadsheets. 'True GIS' (Elshaw Thrall and Thrall 1999) nevertheless remains a distinctive software environment for the conduct of spatial analysis, by virtue of its data management, transformation, and linkage functions, the range of analytical techniques which it supports, and its spatial representation and display functions. Much has been written about the linkage of specialist spatial analysis functions to GIS in ways which are consistent with the linear project design of Fig. 6.1 (Anselin and Getis 1992).

Yet emergent computer technologies are also being harnessed to new computation-intensive techniques (see Openshaw 1998; Openshaw and Openshaw 1997). The emergent field of *geocomputation* (Longley et al 1998) is associated with new and novel ways of using computers to address spatial problems, and may engender new generic approaches to spatial problem solving. To date, as Couclelis (1998) observes, geocomputation has encompassed more or less the full 'grab bag' of spatial analysis techniques set within computationally-intensive environments: yet a common thread to most research applications has been their heavy emphasis upon inductive generalization.

As such, the geocomputational approach would seem set on a collision course with the principles of scientific research epitomized by the linear project design. The development of the approach appears to some to be more a reaction to developments in technology rather than science, although others see it as a deeper quest to develop new science using new tools. This is much more than just an academic or moot point: for some, the new automated 'artistic science of statistical analysis' epitomized by the work of Openshaw (1998) and colleagues is an over-egging of a more mundane reality—namely that geocomputation is essentially reactive, a 'black box' technological response, which is diverting attention from more truly active and rigorous use of GIS tools to develop GI Science (Couclelis 1998; Goodchild 1992). Technology empowers us with tools, so this argument goes, yet conventional wisdom asserts that we need consciously and actively to use them in developing science without surrendering control to the machine. The implication of this view is that better models invariably require better theories, and that these should be transparent and couched within more conventional scientific paradigms (e.g. Burrough 1998). An opposing view is that for too long the esteem in which academics hold traditional deductive science has diverted us from the other equally valid quest for predictive success, and that computationally-intensive inductive generalization is the best vehicle to achieve this. A middle path is suggested by Mandelbrot's (1986) observation that '[d]escription coming before theory is the normal pattern in science', and the view that visualization of pattern creates insight into process (Batty and Longley 1994). Certainly visualization and the creation of virtual realities (VRs) enhances our understanding of scientific problems, and the work of Shiffer (1998) has illustrated the ways in which networked VRs might be used by different users who can each bring their own expertise and insights to collective problem-solving.

From ontology to prediction and practice, debate is likely to characterize the evolution of geocomputation although, as Couclelis (1998) observes, there is no obvious single meta-theoretical framework for activity in prospect. While this state-of-affairs is resonant with recent postmodern fashions, it nevertheless leaves geocomputation with rather looser anchorings than the conventional spatial analysis paradigm. The jury is also still out as to whether the substantive conclusions of such approaches can be sustained under challenge, and whether rich digital descriptions will ever be more than an exploratory prelude to analysis within the more conventional spatial analysis paradigm.

6.3 The changing data infrastructure

Developments in information technology and computing have strong steering influences on the direction and development of spatial analysis research, yet this is only part of the story. However computer models are linked together, all empirical models are ultimately grounded in data. In this section we identify some of the changes to the data environment that are taking place, and use two case studies to illustrate some of the hybrid data modeling methods and techniques that are presently under development. We also speculate on the kinds of spatial analysis that we expect to dominate in the future.

6.3.1 Changes in supply, pricing and access

In physical and social science alike, the costs of data have generally been a (sometimes **the**) major component of the costs of GIS creation. The order of magnitude of data costs reflects a number of technological and secular imperatives which govern the supply, pricing, and access aspects to data availability.

In the early days of GIS, the 'data bottleneck' of (manual or semi-automated) digitizing presented a major impediment to the creation of spatially-referenced databases, particularly if the hard copy source documents were complicated or ambiguous. Early software systems provided (by present day standards) fairly unsophisticated procedures for detecting and correcting the results of error-prone digitizing. Moreover, 'framework' spatial data, such as those created and maintained by national mapping agencies (Rhind 1997) were available only in hard-copy printed form, and in the early days of GIS there was resistance to initiating the task of converting 'legacy' hard copy maps to digital form.

A wealth of digital data has since come into existence. First, and as with computer hardware, new technology is playing an important role. In particular, the wide (selective) availability of global positioning systems (GPS; Lange and Gilbert 1999) makes creation of new digital data sets much more straightforward than hitherto. Second, most national mapping agencies have gradually overcome their initial reluctance to create digital versions of their paper records, while at smaller scales private providers have created a range of digital atlas products. And third, computerized logging of the physical and social environment takes place with ever-increasing frequency, and to ever-greater levels of detail—for example through high-resolution remote sensing of the physical and built environments and the digital encoding of consumer purchasing behavior (through loyalty programs and the development of 'relationship marketing') in the socioeconomic realm.

Yet this has not created a panacea for data modeling. In practice, accurate field recording of data remains an expert task and sound geographical analysis presumes sound data standards (Salgé 1999). Many national mapping agencies (such as Great Britain's Ordnance Survey: Rhind 1997, 1999) have only succeeded in 'going digital' in the face of increasingly stringent public expenditure constraints by recovering vastly increased proportions of their creation and maintenance costs through user charges: the inevitable consequence is a rationing of framework data on an 'ability to pay' basis. Similarly hawkish data pricing regimes may apply to the data products from the new generation of high-resolution satellite sensors (see Barnsley 1999), while high royalty charges dissuade many business users from census data and census data products in some countries (such as the UK). At the same time, governments are reluctant to fund even their traditional linear project design-driven surveys, in view of the apparent tide of information created using new data capture technologies. With respect to the academic realm, the rise of interdisciplinary science is leading to a higher incidence of jointly-funded projects, and the commonplace situation in which the creators of spatial data may be widely separated from some of the communities of end users. As creators and users of data become more and more separated, in space, time, and intellectual tradition, the ability to describe data becomes increasingly critical. The creator must be able to tell the user about methods, accuracy, formats, and all of the details needed to transfer, open, and make effective use of the data. Moreover the user must be able to determine whether a given data sets meets or falls short of requirements. The term 'metadata' is increasingly used for data description; metadata combine the functions of a library's card catalogue, handling instructions on the outside of the data package, and the quality 'seal of approval'. Metadata have become the key to data sharing through clearinghouses (e.g. the National Geospatial Data Clearinghouse, *www.fgdc.gov*; and the Alexandria Digital Library, *alexandria.ucsb.edu*).

The rise of interdisciplinary science is leading to the situation in which the creators of spatial data may be widely separated from some of the communities of end users.

6.3.2 The changing remit and requirements of modeling

The early years of the spatial analysis paradigm were associated with the development of wide-ranging models of physical and social systems (e.g. Batty 1981). The remit of such models was avowedly ambitious, yet on reflection the data infrastructure was not commensurate with the tasks in hand. A number of commentators (Batty 1976; Sayer 1979) have identified reasons for the subsequent demise of large scale socioeconomic modeling activity, although the innovation of GIS has brought with it a renaissance in model-building activity. Moreover, any decline in large-scale modeling of socio-economic systems has been matched by the rapid growth of environmental modeling, much of it coupled with or otherwise making use of GIS (Goodchild et al 1993; Goodchild et al 1996).

The new is quite different from the old, however. Within the socioeconomic realm, Birkin (1996) has described how the current generation of spatial interaction models, for example, seeks only to model limited (in terms of spatial extent, time frame and attribute range) aspects of urban sub-systems. This in part reflects secular trends in all developed societies away from system-wide planning, yet it also reflects a profound reappraisal of what we now consider to be the appropriate domain and capability of analytical models. Today's urban models are much more data-rich in two respects. First, the revolution in the supply and availability of geographical information means that data no longer represent coarse zonal aggregations, and thus that the data model of spatial distributions bears a closer correspondence with reality. Second, the first generation of urban models used data derived exclusively from public sector sources and which were thus restricted to the limited range of variables of interest to officialdom. Whilst such data can be used, singly or in combination, to create crude **indicators** of human behavior and activity patterns, such indicators bear at best a very imperfect correspondence with reality.

Within the socioeconomic realm, the present status of modeling is rather ambiguous. Within academia, disenchantment with urban modeling leaves it as an area of activity with a significantly reduced real share of intellectual activity compared to, say, twenty years ago. Business applications of data-rich partial models of components of urban systems are buoyant, and today client repeat purchases provide vindication of the validity of spatial interaction and other modeling approaches. Within planning, there has never been a greater need for accurate data and analytical models of urban systems, because the rate, scale, and pace of change has never been greater. Yet, in the UK at least, there is disquiet about the 'predict and provide' approach to planning which has hitherto been based upon aggregate modeling approaches.

6.3.3 Model linkage: towards a new perspective?

The linear project design that presented in Section 6.2.1 presumed that resources were available for a linear, vertically integrated sequence of events. Today's research environment is much less straightforward. The strictures of public expenditure make it less likely that large-scale purpose-specific research will be funded, while information commerce makes it less than unequivocal that the best secondary data will be available. Yet data warehouses are bursting with data that might be combined to create richer profiles of landscapes, morphologies, households, and activity patterns than have ever been created before. While the developing geocomputation paradigm presents us with some 'brute force' mechanisms for searching out generalizations from large and complex data sets, we may have no way of knowing whether such generalizations hold any scientific validity.

A negative view of this research environment would suggest that a price has been put on scientific truth that lies beyond the budget of many researchers. There is some truth in this, yet economic imperatives need also to be viewed in their technological context. In truth, as our retrospective of urban modeling above has illustrated, data collected through the linear project design did not provide a panacea in practice. Today's digital data infrastructure is more detailed, relevant, and up-to-date than ever before. The problem is that this infrastructure is also more piecemeal, and hence possibly ill-founded and unsafe.

GIS has always been an applications-led technology, and this is more true today than ever before. The sophistication of current applications requires a breadth and depth of data that could never have been sustained by established data collection methods. Today's open and desk-top GIS alike are geared towards the analysis of application-specific 'horses for courses' data sets. Such data sets are required to model real-world systems that are dynamic and fast-changing, and thus the time scale between data collection and availability of secondary analysis needs also to be shortened. Our understanding of physical and social systems alike is now of such sophistication that infrequently collected, aggregate, and surrogate spatial data are simply not good enough. These are all crucial considerations, yet they all lie outside the remit of the linear project design. Are we therefore faced with a stark choice between scientific validity and 'making do' with inappropriate, overly-aggregate, out-of-date indicators? The rejection of Census-based geodemographics in favor of lifestyles (i.e. data warehouse) analysis in much of business geographics suggests that the road to scientific truth is no simple one-way street, and that proponents of inductive data-led thinking have their supporters in the world of application.

Table 6.1. Some characteristics of UK Census data and satellite imagery

Census	Satellite imagery
Vector data structure, e.g. UK SASPAC comprises attribute files compatible with vector boundary data at enumeration district (ED: c. 160 households) scale	Raster data structure: finer scale of spatial resolution (e.g. SPOT pixel size 30m x 30 m)
Infrequently collected – usually decennial	Frequent repeat passes
Residential building types only: averaged across EDs which are assumed uniform areas	All building types identified but not differentiated, plus spatial distribution of urban infrastructure and public open space
Very limited differentiation between non-residential land use functions	Greater differentiation of land use by inference from land cover
Information on socio-economic characteristics of households/individuals	-
Misrepresentation of fine scale spatial distributions—choropleth maps imply uniform within-area density of attributes	Better representation of detailed spatial form, but poor differentiation of function
Comprehensive coverage in theory—but evasion in practice, concentrated in inner city areas	Comprehensive coverage
Temporal inconsistency of (artificial) boundaries	Consistent, subject to minor variation in sensor characteristics
Long historical time frame for temporal analysis, but only in digital form post 1971	Landsat available post 1984

Framed in these terms, one of the big questions for GIS at the turn of the millennium must be: Can the new digital data infrastructure be assembled together in a sufficiently accurate, orderly and rational way to bridge relevance, richness and academic respectability? Goodchild and Longley (1999) use the term 'concatenation' to describe the integration of two or more different data sources, such that the contents of each are accessible in the product. The polygon overlay operation is one simple form of concatenation. They use the complementary term 'conflation' to describe the range of functions that attempt to overcome differences between data sets, or to merge their contents. Conflation thus attempts to replace two or more versions of the same information with a single version that reflects the pooling, or weighted averaging, of the sources.

> Can the new digital data infrastructure be assembled together in a sufficiently accurate, orderly and rational way to bridge relevance, richness and academic respectability?

6.3.4 Model linkage in practice: RS–GIS concatenation

Census information and satellite imagery are diverse sources of information, and some of their different characteristics (taking the example of UK Census data) are shown in Table 6.1. Longley and Mesev (1997) use information from the 1991 UK small area census statistics as ancillary information to improve the classification accuracy of a contemporary (LANDSAT TM) image of Bristol. Information from the Census is used to assist in sample training and post-classification sorting. The full procedure is illustrated in Fig. 6.2. The resultant hybridized data set is designed with a specialized purpose in mind—to provide detailed data models of the distribution of population and domestic property. This is used to reappraise conventional analysis of the density at which urban space is occupied—and through comparisons Longley and Mesev (1997) develop density gradient profiles for different categories of urban space filling, such as 'built form', 'residential', 'households', and 'population'. They demonstrate that the differences between these apparently similar categories are more than semantic, and can heavily condition whether and to what extent we might consider density profiles characteristic of particular settlement types. The optimistic message of this work is that, once the differences between different conceptions of 'urbanity' have been clearly grasped, it is possible to develop a range of customized indicators of urban morphology. In this way, customized GIS-based data models are informing our thinking about the ways in which urban settlements fill space, as well as providing detailed information as to the morphology of particular settlement structures.

6.3.5 Model linkage in practice: conflating geodemographics and lifestyles

'Lifestyles' is a broad term that has been used to describe data pertaining to the consumption of a wide range of goods and services by identifiable individuals and households. Lifestyles data originate from a diverse range of sources, such as guarantee card returns, questionnaires attached to nationally circulated prize draw entries, and market research surveys. They are usually georeferenced through the postcode system (e.g. in the UK to the unit postcode, which typically comprises 15 or so addresses in urban areas). At least one UK 'data warehouse' estimates that it holds up-to-date information on 11 million UK households. Such data have evident use for direct marketing, for past consumption habits are key guides to future behavior. Harris (1999) has analyzed the anonymized individual/household records from one particular lifestyles questionnaire which was mailed out in October 1996. The number of

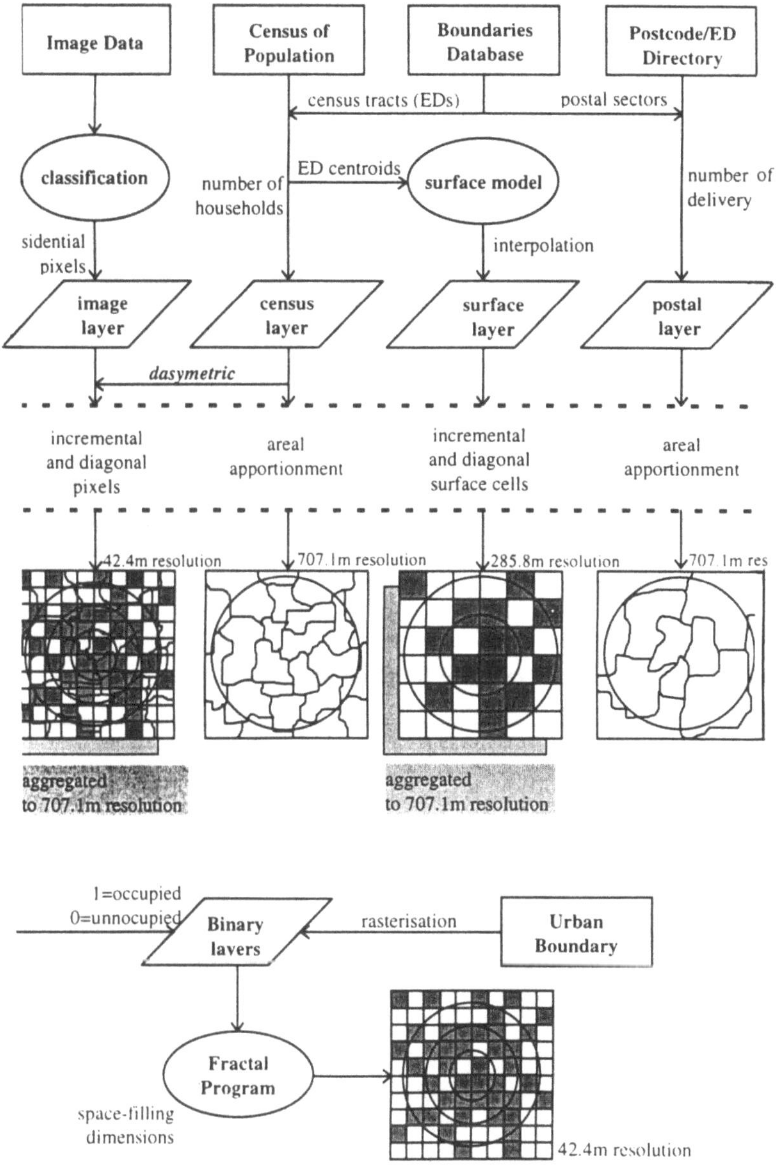

Fig. 6.2 Use of information from the UK Census to assist in sample training and post-classification sorting

respondents to this survey constitutes 10.8% of all households in Bristol, UK (population 636,000): this makes the survey larger in size than a mini census, yet the characteristics of non-respondents are likely to be very unrepresentative of respondents. In recent years, lifestyles approaches have gained some ground as tools for geomarketing at the expense of the use of census and composite geodemographic indicators, because the latter are increasingly out of date (the last UK Census was held in 1991), they are expensive to use because of UK royalty structures and, perhaps most damning of all, the census contains too few variables that bear an identifiable correspondence with consumer behavior (most notably in the UK, because of the absence of an income question in the Census). Some characteristics of lifestyles data vis à vis conventional census-based geodemographics are shown in Table 6.2.

The 'geodemographics–lifestyles' debate thus epitomizes the tensions described in Section 6.3.3 above. Geodemographics is based on tried and trusted techniques and derives from a data set (the Census) which has been designed and implemented using the most rigorous research design principles; and yet at the end of the day, it is out of date, and can supply at best only very imperfect indicators of real-world consumer behavior. Sampling theory tells us that reweighting of largely self-selecting samples on the basis of sub-group response rates is foolhardy; yet survey research practice tells us that quantitative indicators should be direct and transparent, and that survey results are only directly applicable to the population from which the respondents were drawn

Table 6.2. Some characteristics of geodemographics and lifestyles data

Geodemographics (UK)	Lifestyles (UK)
Population (Census) coverage	11 million of roughly 24 million households; little is known about response rates and response rate variability
1991 snapshot data	Continuous updates from waves of mailings
Individual variables of marginal relevance to consumption; no income question and conception of social class arguably outdated	Direct indicators of individual and household consumption
Available at ED level of aggregation	Analysis possible using anonymized individual records at postcode scale of analysis
Constrained to Census geography; integration with postcoded data requires further modeling assumptions	Potential direct application to all other postcoded data
Geodemographic indicators created through application of tried and tested cluster analysis routines	No obvious means of combining unweighted data in a scientifically valid way

(few of us would wholly identify with our digital past-selves who filled out a census form at the start of this decade).

A middle path between these two lies in Batey and Brown's (1995) assertion that lifestyle descriptors can be used as a wrapper to add depth to the labels assigned to different geodemographic groups. Thus, for example, the SuperProfiles category 'affluent achievers' has fairly distinctive Census characteristics in terms of house construction type, socio-economic status and car ownership, to which lifestyle labels about theatre and restaurant patronage, share registers, newspaper readership, and credit card usage are added. The data from which these labels are obtained are in many cases collected by unscientific means or strictly pertain only to coarser aggregations of households. Yet Harris's (1999) cluster analysis of (unweighted) lifestyle data finds some practical validity to this approach: it nevertheless runs rough-shod over conventional views about how scale and aggregation issues should be tackled.

6.4 The future of spatial analysis

Box 6.1 (based upon Goodchild and Longley 1999) summarizes the implications of some of the arguments presented in the preceding discussion, in terms of the kinds of spatial analysis that are likely to become most common in the coming years.

Box 6.1 The implications in terms of the kinds of spatial analysis that are likely to become most common in the coming years.

- analysis of data whose meanings are clearly understood, making it easier for multidisciplinary teams to collaborate;
- analysis of data which are routinely collected in the day-to-day functioning of society and the everyday interactions between humans and computers;
- analysis of data with widespread use, generating demands that can justify the costs of creation and maintenance;
- analysis of data with commercial as well as scientific and problem-solving value, allowing costs to be shared across many sectors;
- methods of analysis with commercial applications, making it more likely that such methods will be implemented in widely available form;
- methods implemented using general standards, allowing them to be linked to other methods using common standards and protocols.

(source: after Goodchild and Longley 1999)

6.5 Concluding comments

Our discussion has highlighted the way in which the advanced information economy of the late 1990s has multiplied the number of potential sources of (rich) digital information, yet in ways which will be less standardized and project-specific than those implied by the linear project design. A major challenge to the GIS community is to devise methods to reconcile diverse data sets with different data structures or spatial referencing systems. Only in this way will GIS be able to tease out the complex relationships that exist between projects, data sets, and analytic techniques in modern science. The self-perception of rigor amongst spatial analysts has hitherto been misplaced because of the vagaries and inadequacies of data quality, resolution and richness: progress requires us to face up to the fact that the linear project design was never a panacea in practice. As Goodchild and Longley (1999) state 'Tomorrow's science will be increasingly driven by complex interactions, as data become increasingly commodified, technology increasingly indispensable to science, and conclusions increasingly consensual. New philosophies of science that reflect today's realities are already overdue.' In the short space of three decades we have moved from a world in which computers were seen as mechanical aids to calculation, to one in which digital information and computing are indispensable to interdisciplinary science. It is still far too early to foresee all of the profound changes that this will bring.

Exercises

1. What is the invariance property mentioned in the introduction? Does it provide a robust test of whether analysis is spatial or not?
2. Consider one of the global radiation models from Chapter 2. Under what circumstances would a GIS be suitable for modeling? What would be the benefits, and what the requirements? Consider all the elements of data, objects, research, and results.
3. Consider the problem of integration of data and suppose that a researcher wishes to apply GIS to the pea model (Chapter 1). Discuss the problems that might be encountered, and give some solutions.
4. Scientists are trained never to approach analysis as a 'black box', but always to understand all of the stages of manipulation of their data. Yet modern computing technology seems to leave the user more and more in the dark about the details of processing. Is this a problem, and how is it likely to be resolved?

7 Multi-scale approaches for geodata

M Molenaar

7.1 Introduction[1]

7.1.1 Four strategies

Topological object relationships in combination with object classification hierarchies appear to be fundamental in the definition of the aggregation rules for spatial objects. Such rules are essential building blocks for the construction of generalization procedures in spatial databases. A model for spatial database generalization can be formulated based on the syntax of the Formal Data Structure (FDS) as proposed in (Molenaar 1989). The syntax of the FDS will be formalized first, then database generalization procedures will be formulated with this syntax.
Four strategies will be explained for generalization:

- *geometry driven generalization*, the change of geometric resolution cells determines the transition from entities at a large scale to new entities at a smaller scale,
- *class driven generalization*, spatial objects at a large scale forming a region under one thematic class are merged for representation at a smaller scale,
- *functional generalization* links objects that are considered as response units in processes defined at different scale levels,
- *structural generalization* gives a stepwise simplification of a spatial process description in an area.

These different strategies will be explained and compared. In the discussion spatial data generalization will be presented as data transformation processes.

7.1.2 Spatial processes at multi-scale levels

Multi-scale approaches are at present a focal point of GIS research because awareness is rising that many processes on the earth surface can only be monitored and managed if they are understood in their geographical context. Part of this context is defined by the scale range at which these processes work. The development of land use in a district, for example is driven by actors at a lower aggregation level such as farmers, residents and companies. Their activities are constrained by socio-economic conditions and infrastructure at regional level and by the macro economic planning at national and supra national level. Another example is development of natural vegetation cover. Its actual state is defined by co-occurrence of species forming vegetation types, which are part of eco-systems. Its development is constrained by climatic conditions, geologic and soil condition and hydrology. Here too we find hierarchical levels of organization.

---

[1] The content of this chapter has been based on the concepts precribed in Molenaar (1998)

Monitoring and management of multi-scale processes requires information at different scale levels. A research problem for GIS is:

- to decide which information should be handled at each scale level for each type of process,
- to develop methods for transferring information between the different scale levels so that duplication of expensive data acquisition is avoided as much as possible, and consistency between data at different scale levels is maintained.

This second item is strongly related to map generalization. Multi-scale approaches in a GIS environment, however, can be also dealt with by data base generalization operations. These are more flexible and allow a different approach.

Multi-scale approaches derive both terminology and many concepts from cartography and object oriented computer science (Kemp 1990). This mixture of different idioms often leads to confusion and fuzzy terminology. In this chapter a data base perspective for multi-scale approaches is presented, emphasizing topologic and semantic (hierarchical) data models. Different concepts that play a role in the generalization of spatial data will be discussed in relation to strategies to be used for solving multi-scale problems.

7.1.3 Spatial databases and multi-scale problems

A spatial database contains data representing elementary statements about a spatial situation. These elementary statements refer to relationships between objects, geometric data, thematic data etc. Query operations may derive more relevant information for the user, e.g. about the state of the objects and about their mutual relationships. The semantics of derived statements is generally of a higher level of complexity than the stored data. They should help the user to understand the structure of a mapped area. They often refer to spatial relationships between the mapped objects. If the area structure should be understood at a higher abstraction level though, these derived statements could also refer to relationships among aggregated objects. The understanding of the structure of an area at several abstraction levels is strongly related to the problem of spatial generalization and multi scale representations.

Aggregation hierarchies for spatial objects can serve as basic tools for multiple representations of geo-data within the context of conceptual generalization processes (information abstraction). These aggregation hierarchies can be based on the formal data structure (FDS) for single valued vector maps (Molenaar 1989, 1998) to combine aspects of object-oriented and topologic data models. Point-, line- and area objects are represented with their geometric and thematic data. The geometric representation supports the analysis of topologic object relationships, whereas the thematic description is structured in object classes that form generalization hierarchies. These class hierarchies together with topologic object relationships support definition of aggregation hierarchies of objects. Classification and aggregation hierarchies play an important role in linking the definition of spatial objects at several scale levels. These structures are fundamental in the definition of rules for modeling generalization of spatial information at different resolution levels. The capacity of Geographical Information Systems (GIS) to register and handle topological information in combination with object hierarchies makes them very useful for automation of conceptual generalization of spatial data.

In a cartographic context, generalization can be defined as abstracting the representation of geographic information when the scale of a map is changed. It involves abstraction of both thematic and geometric object data. We will distinguish two phases:

- a conceptual generalization phase, which implies determination of representation content in the generalized situation and definition of rules how generalized objects can be derived from the objects at a lower generalization level
- a graphical generalization phase which applies algorithms for geometric simplification of shapes and symbols to assure map legibility.

Information abstraction is mainly determined by expert knowledge and can usually be expressed as logical rules. These rules are susceptible to be translated as database management procedures in a GIS (Martinez Casasnovas 1994; Richardson 1993). Several processes are recognized for information abstraction: *classification*, *association*, *(class) generalization* and *aggregation*. Class generalization and aggregation are directly related to changes in the level of definition of objects when the mapping scale changes. Aggregation is the combination of elementary objects to build composite objects and will be based on two types of rules:

- rules specifying classes of elementary objects building a composite object;
- rules specifying geometric characteristics such as minimum size and topological relationships of these elementary objects (i.e. adjacency, connectivity, proximity, etc.).

The syntactic structure of a data model for handling topologic and hierarchical relationships between spatial objects will be explained in this article. Processes for database generalization will be formulated with this data model.

7.2 A spatial data model for multi-scale approaches

7.2.1 Topologic structures for the representation of spatial objects

Entity types for spatial data

The spatial structure of an area can be expressed in terms of point-, line- and area objects. Their spatial extend and their topologic relationships will be expressed by means of a set of geometric elements. Frank and Kuhn (1986) showed that the geometric structure of a vector map can be described by cell complexes. For a two-dimensional map these consist of 0-cells, 1-cells and 2-cells. The 0-cells and 1-cells play a similar roles as nodes and edges when the geometry of the map is interpreted as a planar graph. The 2-cells can then be compared to the faces related to the planar graph through Eulers formula (Gersting 1993). The terminology of the planar graph interpretation will be used here, but their relationships will be formulated as those for cells according to the concepts presented in Molenaar (1994, 1998). This formulation is then based on

- three geometric types: *nodes* ($\mathtt{n}$), *edges* ($\mathtt{e}$) *and faces* ($\mathtt{f}$).
- three geometric object types: *point objects* (O_p), *line objects* (O_l), *area objects* (O_a)

The reader will recognize that the formalization explained in this paper is to a large extent isomorphic with topologic data structures defined for GIS such as ATKIS/DLM, DGF, TIGER and DIME (Hesse 1992; Walter 1994; Marx 1990; USBUREAU 1990). This formalization will be based on the FDS described in Molenaar (1989, 1998).

Relationships between nodes, edges and faces

The following relationships can be defined between geometric elements of a planar graph.

Edge e_i has node n_j as its begin node	$\rightarrow$ $\text{Begin}[e_i, n_j]=1$ otherwise=0
Edge e_i has node n_k as its end node	$\rightarrow$ $\text{End}[e_i, n_k]=1$ otherwise=0

We will consider edges as straight line segments. Each edge will always have one face at its left hand side and on at its right hand side. These relationships will be expressed by the following functions

Edge e_i has face f_a at its left-hand side $\rightarrow$ $\text{Le}[e_i, f_a]=1$

If $f_b \neq f_a$: $\text{Le}[e_i, f_b]=0$

Edge e_j has face f_a at its right-hand side $\rightarrow$ $\text{Ri}[e_j, f_a]=1$

If $f_b \neq f_a$: $\text{Ri}[e_j, f_b]=0$

If edge e_i has face f_r at the right hand side and face f_l at the other side then these faces are adjacent at this edge which will be expressed by the function

$\text{ADJACENT}[f_r, f_l | e_i]=1$ (and =0 otherwise)

$\text{ADJACENT}[f_r, f_l]=1$ (and =0 otherwise)

Line objects

The geometry of a simple line object O_l is represented by a chain of edges as in Fig. 7.1a. If edge e_p is part of O_l then this can be established by the function $\text{Part}_{11}[e_p, O_l]$. The notation $\text{Part}_{uv}[\ ,\ ]$ means that an entity with spatial dimension u is a part of an entity with dimension v. If the edge is part of the object then $\text{Part}_{11}[\ ,\ ]=1$, else $=0$. A line object O_l will have a begin node $n_b=\text{Begin}(O_l)$ and an end node $n_e=\text{End}(O_l)$. These can be found through the edges of O_l. The direction of O_l can then be specified by $\text{Dir}[O_l]=\{n_b, n_e\}$.

Area objects

The geometry of a simple area object O_a is represented by one or more adjacent faces (Fig 7.1b). If face f_v is part of O_a then this will be represented by $\text{Part}_{22}[f_v, O_a]=1$. The set of all objects from which a face f is a part is $\text{OA}(f)=\{O | \text{Part}_{22}[f, O]=1\}$. The function $\text{OA}(\ )$ relates two two-dimensional entities. For overlapping area objects a

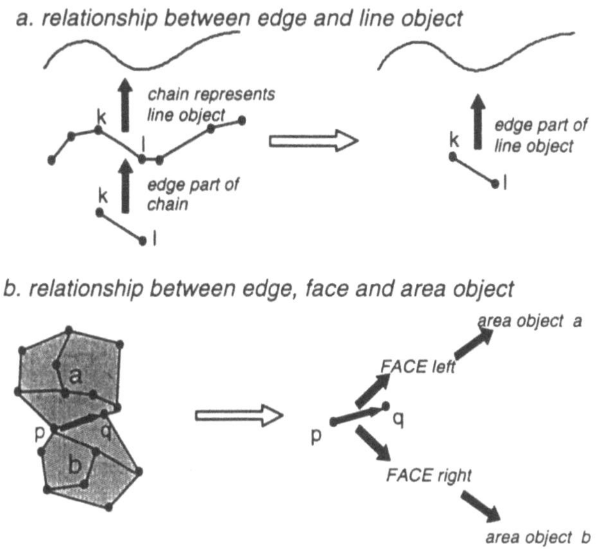

Figure 7.1: Relationships between edges and objects.

face might be part of several objects, but each object will also consist of one or more faces. Therefore this is a many-to-many relationship. Overlapping objects can be found through their common faces.

It is now possible to check whether edge e_i is related through face f_v to an area object O_a There is at most one face for which both $Le[e_i,f_v]=1$ and $Part_{22}[f_v,O_a]=1$. If such a face exists then the function relating the edge to the object will be equal to 1, in all other cases to 0. Hence if e_i has area object O_a at its left-hand side then $Le[e_i,O_a]=1$ else $=0$. Similarly if edge e_i has area object O_a at its right-hand side then $Ri[e_i,O_a]=1$ else $=0$. Combination of these two functions gives for edge e_i:

$$B[e_i,O_a]=Le[e_i,O_a]+Ri[e_i,O_a]$$

If edge e_i is part of the boundary of O_a then only one of the functions $Ri[e_i,O_a]$ and $Le[e_i,O_a]$ is equal to 1 and hence $B[e_i,O_a]=1$. If e_i has O_a both at its left-hand side and at its right-hand side then it is running through O_a and $B[e_i,O_a]=2$. If $B[e_i,O_a]=0$ there is no direct relationship between e_i and O_a.

Adjacent area objects

When an edge has object O_a at its left and not at its right and object O_b at its right and not at its left then O_a and O_b are adjacent at this edge. If the objects do not overlap, i.e. they have no common faces and they are adjacent at least at one edge, then they are adjacent: $ADJACENT[O_a,O_b]=1$ (and $=0$ otherwise).

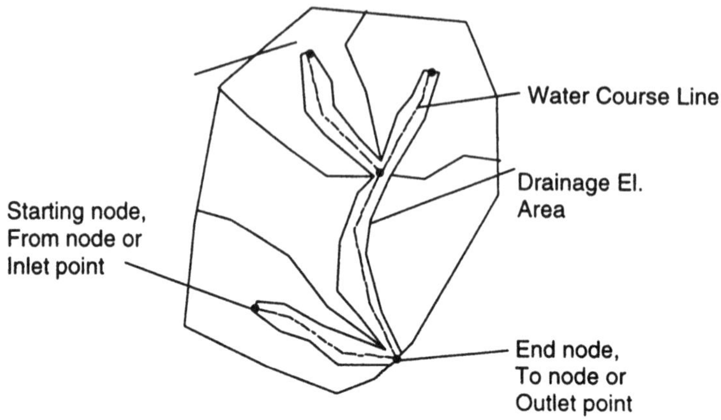

Figure 7.2: Elementary objects in a drainage system

Line- and area objects

Several relationships between a line object O_l and an area object O_a can be found by checking for each edge of O_l how it is related to O_a This will be expressed by the functions

$$Le[O_l, O_a | e_i] = Min(Le[e_i, O_a], Part_{ll}[e_i, O_l])$$
$$Ri[O_l, O_a | e_i] = Min(Ri[e_i, O_a], Part_{ll}[e_i, O_l])$$

The function $Le[O_l, O_a | e_i]$ equals 0 if O_a is either not at the left of e_i, or e_i is not a part of O_l. It equals 1 if e_i is both at the left of O_a and is a part of O_l. It equals 2 if O_l runs through O_a with edge e_i. A similar explanation applies to the function $Re[O_l, O_a | e_i |]$.

For the relationship between a line object O_l and an area object O_a we can write

$$B[O_l, O_a | e_i] = Le[O_l, O_a | e_i] + Ri[O_l, O_a | e_i]$$

If $B[O_l, O_a | e_i] = 2$ then O_l runs through O_a at edge e_i. If $B[O_l, O_a | e_i] = 1$ then it is at the border and if it equals 0 then there is no relationship. The relationship between the two objects might be different at different edges.

A hydrologic example

For modeling hydrological systems three types of elementary objects are defined according to (Martinez Casasnovas 1994): water course lines, drainage elements and catchments (Fig. 7.2). Drainage elements are gullies, each element has a catchment area from which it receives overland flow of water. Each element also receives water from upstream elements (if there are any) and empties into a downstream element. The water flow through each element is represented by a water course line.

The relationship between these objects is 1-1: each drainage element D_i contains exactly one water course line W_i and is embedded in exactly one sub-catchment area C_i.

A sub-catchment area may be dissected by its drainage element (Fig. 7.2), but is still considered as one sub-catchment. The topologic relationships between these objects can be expressed by functions of section 7.2. For water course line W_i and drainage element D_i:

$$(\forall e_k | \mathrm{Part}_{11}[e_k, W_i]=1) \Rightarrow B[W_i, D_i | e_k]=2.$$

This will be written shortly as $B[W_i, D_i]=2$. For $j \neq i$ $B[W_i, D_j]=0$. It means that W_i runs through D_i so that it has D_i at both sides and it is not related to any other drainage element. This is a topologic restriction due to a semantic constraint from the context of the hydrologic model. Another semantic constraint is: for drainage element D_i and catchment C_i $\mathrm{ADJACENT}[D_i, C_i]=1$, whereas $\mathrm{ADJACENT}[D_i, C_j]=0$ for $j \neq i$, so that D_i is only adjacent to C_i and to no other catchment.

Each drainage element is also connected to a downstream element and, depending on its position in the network, to one or more upstream elements. The relationship between drainage elements can also be found through the watercourse elements. These should be directed according to the direction of the water flow. For each W_i we can find the upstream element W_h through the rule $\mathrm{End}(W_h)=\mathrm{Begin}(W_i)$. This relation will be expressed by $\mathrm{Upstr}[W_i, W_h]=1$, whereas $\mathrm{Upstr}[W_i, W_j]=0$ otherwise. Due to the 1-1 relationships between W, D and C the upstream relationship can be transferred as follows

$$\mathrm{Upstr}[W_j, W_i]=\mathrm{Upstr}[D_j, D_i]=\mathrm{Upstr}[C_j, C_i]$$

so that the order relationships between the water course lines can be translated into order relationships between the areas in which they are contained. We will assume here that the stream network structure is defined so that for each W_j with a Strahler number[2] > 1 there are two or more upstream water lines W_i, but for each W_i there is only one downstream water line W_j.

7.2.2 Object classes and class hierarchies

Terrain objects refer to features on the earth surface. In a systems environment they are interpreted with a thematic and geometric description. In most applications terrain objects will be grouped into several distinct classes with a list of attributes connected to each class (Molenaar 1998). Let C_i be a class, and let the list of its attributes be $\mathrm{LIST}(C_i)=\{A_1, A_2, \ldots\ldots, A_n\}$. Attribute lists will be different for different classes, i.e. for $i \neq j$ $\mathrm{LIST}(C_i) \neq \mathrm{LIST}(C_j)$. Terrain objects inherit the attribute structure from their class. Each object therefore has a list with a value for each class attribute, thus for member c of class C:

$$\mathrm{LIST}(c)=\{a_1, a_2, \ldots\ldots, a_n\}$$

---

[2] The Strahler number specifies the place in the hierarchy of an object. Simple objects get Strahler number 1, aggregated simple objects get Strahler number 2, etc.

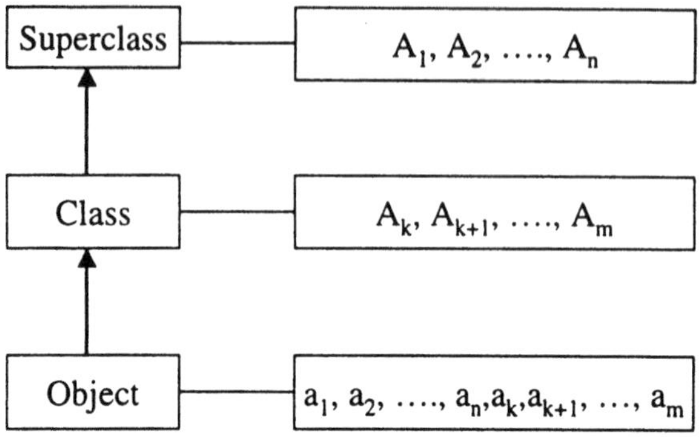

Figure 7.3. The hierarchical relationships between objects and classes and their attributes.

where $a_k = A_k(c)$ is value of A_k for object c for all $c \in C$ and $A_k \in LIST(c)$. When two or more classes have attributes in common, then a superclass can be defined with a list containing these common attributes as superclass-attributes (Molenaar 1993, 1998). The original classes are subordinated to these super classes, for example, the class *forest* is a superclass containing subclasses such as *deciduous*, *evergreen*, and *mixed forest*. The terrain objects are then assigned to these classes.

With these observations we find a class hierarchical structure (Fig. 7.3). In literature on semantic modeling (Brodie 1984; Brodie et al 1984) the upward links of the classification hierarchy are labeled respectively as [IS A] links. These links relate each particular object to a class and to super classes.

It is possible to add more hierarchical levels to the structure. At each level the classes inherit the attribute structure of their superclass at the next higher level and propagate it normally with an extension to the next lower level. At the lowest level in the hierarchy are the terrain objects, at this level the attribute structure is not extended any more, but here inherited attributes are evaluated. In this case we find for e:

$$LIST(e) = \{a_1, a_2, \ldots, a_n\}$$

where $a_1 = A_1(e)$ is value of A_1, $A_1 \in LIST(c) \cup LIST(SC) \cup \ldots$.. Thus A_1 is an attribute of the *class* or *superclass(es)* of e. If the classes at each level are disjoint so that the hierarchy has a tree structure then the terrain objects will get their attribute structure only through one inheritance line in the hierarchy, i.e. they have a unique thematic description. We will work under this assumption in this chapter.

Terrain objects occur at the lowest level in the classification hierarchy. They can be seen as the elementary objects within the thematic field represented by the classification system. This implies that the decision, whether certain terrain objects should be considered as elementary or not, should always be made within the frame work of a thematic field. Objects that are considered as elementary in one thematic field are, however, not necessarily elementary in another thematic field.

7.2.3 Object aggregation

Objects can be aggregated to build composite objects at several levels of complexity, forming aggregation hierarchies. An aggregation hierarchy shows how composite objects can be built from elementary objects and how these composite objects can be put together to build more complex objects and so on (Brodie 1984; Brodie et al1984). The upward relationships of an aggregation hierarchy are called PARTOF links. These links relate a set of objects to a composite object, a more complex object, and so on. For example, 'Field A is PARTOF John's farm is PARTOF the Meuse catchment.'

For composite spatial objects PARTOF links might be based on two types of rules involving thematic and geometric aspects of elementary objects. Consequently the generic definition of an aggregation type should consist of the following rules (Molenaar 1993, 1998):

- rules specifying classes of elementary objects building an aggregated object of this type,
- rules specifying geometric and topologic relationships among these elementary objects.

Suppose that we wish to form aggregated objects of type T. To do so we first identify the objects O_i that could be part of such aggregates. These objects must fulfil certain criteria, which according to the two sets of rules given earlier will be based on thematic data of the objects. Let these criteria be expressed by a decision function $D(O_i, T)=1$ if the object fulfils the criteria and $=0$ otherwise.

Regions can now be formed by applying two rules:

all objects O_i in region R_r satisfy the decision function for T

$$(\forall O_i \mid O_i \in R_r) \Rightarrow D(O_i, T)=1$$

All objects O_i that satisfy the decision function for T and that are adjacent to objects of the region belong to the region

$$(\forall O_i \mid D(O_i, T)=1)(\exists O_j \in R_r \mid \text{ADJACENT}[O_i, O_j]=1) \Rightarrow (O_i \in R_r)$$

The second rule implies that a region can be formed when at least one object has been identified that fulfils the first rule. This object is then the seed around which the region can grow by identification of the other objects that fulfil both rules.

A region R_r can be expressed as a set of objects, i.e. $R_r=\{\ldots., O_i, \ldots.\}$. The objects of the region can be aggregated to form an aggregated or composite object O_{ar}, the suffixes express that the object is of aggregation type a and r is its identification number. The operation will be expressed by

$$O_{ar}=\text{AGGR}(R_r)=\text{AGGR}(\{\ldots., O_i, \ldots.\})$$

The fact that O_i is part of O_{ar} is expressed by

$$\text{Part}_{kl}[O_i, O_{ar}]=1$$

where the indices k and l refer to the dimension of the objects. The reverse relation expresses that the object O_{ar} consists of the region R_r, i.e. the function identifies the object that are the components of O_{ar}:

$$\begin{aligned} \mathrm{COMP}(O_{ar}) &= R_r = \{\ldots\ldots, O_i, \ldots\ldots\} \\ &= \{O_i \mid \mathrm{Part}_{kl}[O_i, O_{ar}]=1\} \end{aligned}$$

The geometry of the aggregates can be found through the geometry of the original objects, for each geometric element we can check whether it will be part of an aggregated object of type T_a. This should be done in two steps, which will be explained for the faces of an area object O_i in relation to an aggregated area object O_{ar} The first step evaluates the function:

$$\mathrm{Part}_{22}[f_j, O_{ar} \mid O_i] = \mathrm{Min}(\mathrm{Part}_{22}[f_j, O_i], \mathrm{Part}_{22}[O_i, O_{ar}])$$

this function expresses whether the face is related to an aggregate through object O_i. If that is true then both functions in the expression at the right hand side of the equation are equal to 1 and hence the function at the left hand side as well. Otherwise, at least one of the functions at the left hand side equals 0, and so also the function at the left hand side.

The second step is the evaluation of

$$\mathrm{Part}_{22}\ [f_j, O_{ar}] = \mathrm{Max}_{Oi}(\mathrm{Part}_{22}[f_j, O_{ar} \mid O_i])$$

If there is any object through which the face will be part of an aggregate then this function will equal 1, otherwise it will equal 0. If this function has been evaluated for all map faces then the geometry of the object O_{ar} can be found through their adjacency graph. For edges e_i of these faces the boundary function $B[e_i, O_a]$ can be evaluated and with this function the boundary edges can be found (i.e. $B[e, O]=1$) and through these the topologic relationships with the other objects.

The geometry of the aggregated area object O_a can sometimes be simplified by a reduction of the number of faces. Therefor the edges e_i should be identified for which $B[e_i, O_a]=2$, i.e. the interior edges. If these edges are not part of some line object so that $LO(e_i)=\varnothing$ then they do not carry any semantic information at this aggregation level and could therefore be eliminated.

So far we considered situations where a face is related through an area object to an aggregated are object, so that all involved elements are of dimension 2. Other combinations of dimensions might occur as well. This could be the case if, for example, an edge is related through a line object to an aggregated area object, e.g. it is related through a river to a country.

It is possible to define aggregation types by means of their construction rules. If elementary objects are combined to form a compound object, their attribute values are often aggregated as well. We will see in section 7.3.3 that farm yield is the sum of the yields per field, and that yield per district is the sum of farm yields. Desaggregation of such values is usually difficult because an aggregation hierarchy has a bottom-up character, in the sense that the elementary objects from the lowest level are combined to compose increasingly complex objects as one ascends in the hierarchy. The compound objects inherit attribute values from objects by which they are composed.

The PARTOF relations connect groups of objects with a certain aggregate and possibly on a higher level with another even more complex aggregate, and so on. That means that an aggregation hierarchy expresses the relationship between a specific aggregated object and its constituent parts at different levels. This differs from class hierarchies where classes at several generalization levels can be defined with their attribute structure, but where objects can be assigned to these classes in a later stage of a mapping process.

7.3 Strategies for object generalization

The formalism of the previous chapter helps us to express the structure of spatial data sets. This can be done in an abstracted sense, i.e. without any reference to the logic model of any implemented spatial data base. The four basic operations that will be used in generalization are:

- *selection* of objects to be represented at the reduced scale, this selection will be based on the attribute data of the objects,
- *elimination* from the data base of objects that should not be represented,
- *aggregation* of area objects that should not be represented individually,
- *reclassification* of the generalized objects.

For these four operations information about the spatial structure of the mapped area will be required. Firstly to check which relationships the objects have with their environment and what the effect of their eventual elimination will be on the spatial structure of that environment. Secondly this information is required to formulate aggregation rules for the objects that are to be merged. Once the process has been formulated one can choose how to implement it in any suitable database environment. The hydrologic example in sections 7.2.1 and 7.3.4 has been implemented in an Arc\Info environment, but similar applications exist in Oracle and Prolog.

Several strategies for database generalization can be formulated with this formalism and these basic database operations (Molenaar 1998). These are:

- *geometry driven generalization*: in this strategy it is the geometric information that drives the aggregation process. An example is a raster structure for the geometry of the spatial data. If raster resolution is decreased, i.e. when cell size increases, then the original, smaller cells are merged into new larger cells. The thematic information carried by the original cells should then be transferred to the new cell.
- *class driven generalization*: in this strategy regions are identified, consisting of mutually adjacent objects belonging to the same class. These objects are aggregated to form larger spatial units with uniform thematic characteristics. Generalization is driven by thematic information of the spatial data.
- *functional generalization*: spatial objects at a low aggregation level are aggregated to form new objects at a higher level. Objects are functional units with respect to a

process at the aggregation level and processes at different aggregation levels are related.

- *structural generalization*: the main aim of structural generalization is to simplify description of a spatial system, such as drainage networks, but keeping the overall structure intact: after generalization the total functioning of the same system can be understood at a less detailed level.

Each strategy has its own range of applications. Data base users should be well aware of why they are generalizing spatial data, so that they can chose which strategy is to be used. The first strategy is in most cases used when the geometric resolution of a spatial description is reduced without a clear semantic motivation. The latter three strategies, however, are semantically defined and motivated. They will be explained in some more detail now.

7.3.1 Class driven object generalization

Suppose that a database contains a detailed description of a terrain situation with agricultural fields, forest areas and natural grasslands (Fig. 7.4a). This description might be too detailed for a structural analysis on areas covered by different major types of land use and their spatial distribution. A less detailed spatial description is obtained, if the original objects are aggregated to form larger spatial regions per major land use class.

Figs. 7.4 and 7.5 show that this less detailed description can be obtained in two steps:

1. fields are assigned to more general classes representing the major land use types (Fig. 7.4B);

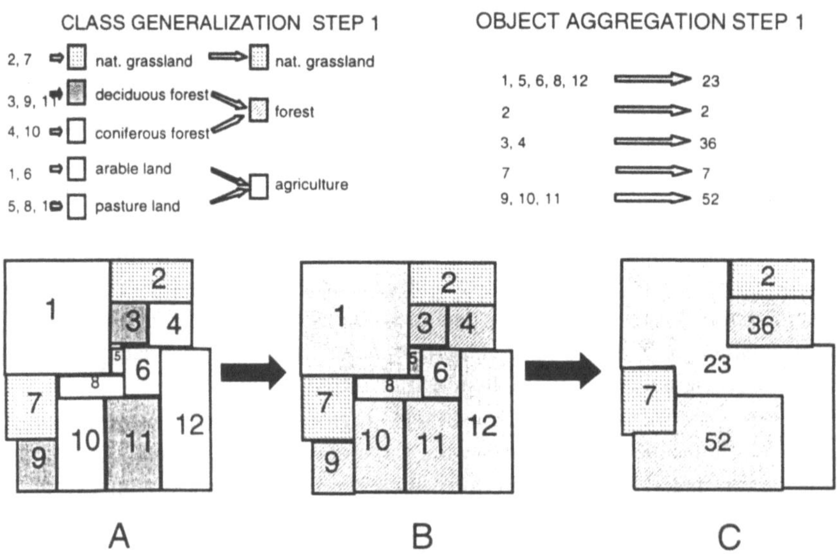

Figure 7.4 Class driven object aggregation.

2 mutually adjacent objects are combined to form regions (Fig. 7.4C).

These final regions are aggregated objects. The functions D(O,T) express then that objects should be aggregated per (super)class, i.e. if aggregated objects should be formed for agriculture then

```
if O∈Agriculture   then D(O,Agriculture)=1
                   else D(O,Agriculture)=0.
```

The output of the aggregation process are regions in the sense of section 7.2.3. Each region is an aggregate of objects that belong to one land use class, so if R_a is an agricultural region then:

for all objects $O_i \in R_a$ is $D(O_i, Agriculture)=1$

if $O_i \in R_a$ and $ADJACENT[O_j, O_i]=1$ and $D(O_j, Agriculture)=1$
then $O_j \in R_a$

A consequence of this rule is that after the aggregation process there can be no two adjacent regions that are of the same type, i.e. that represent the same land use class.

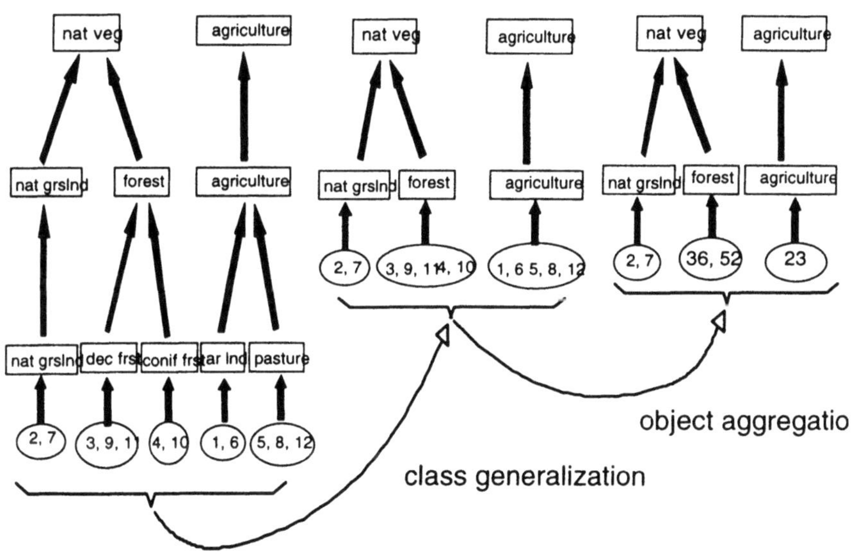

Figure 7.5 A diagram representing the generalization and aggregation steps of the object generalization process of Figure 7.4.

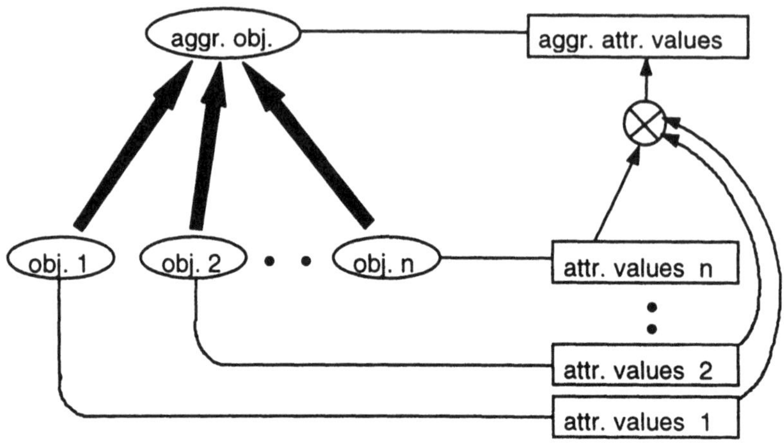

Figure 7.6 The aggregation of attribute values.

7.3.2 Thematic and geometric resolution

The example represented in Figs. 7.4 and 7.5 shows how thematic aspects of newly aggregated objects can still be handled within the original class hierarchy. It might be that the same classes can be used as for the original objects, but the example shows a situation where with each database generalization step the class hierarchy is adjusted. At every step the lowest level of classes is removed, only the more general classes remain (see also Figs. 7.5 and 7.12). That means that thematic resolution is adjusted to geometric resolution of the terrain description.

Sometimes it is not necessary to jump to more general classes with each aggregation step. In those cases new objects can be assigned to the original classes and hence have the same attributes as the objects from which they have been composed. In Fig. 4B, for example, objects 1, 5, 6, 8 and 12 all belong to the class "agriculture" and hence have the same values. Within this class they are aggregated to form the composite object 23, i.e.

$$O_{23}=AGGR(O_1, O_5, O_6, O_8, O_{12}).$$

attribute structure. They were distinct because of different attribute values. This new object still belongs to the same class "agriculture" with attributes in common with the original objects. The attribute values of the original objects are then transferred to the new object as in Fig. 7.6.

When the attribute values are of the ratio scale type then the aggregated value can often be obtained by summation or by taken the average value over the objects that compose the new object, e.g.:

$$A[O_a] = \Sigma_{Oi \in COMP(Oi)} A[O_i]$$

Examples are attributes like wood volume and crop yield and population. For other attributes like vegetation cover or population density it might be that (weighted) averages should be computed.

7.3.3 Functional object generalization

Functional object generalization will always reduce spatial variability of attribute values, because after each aggregation step the attribute values that were assigned per object will be merged into one value for a larger object. This means that relations between spatial and thematic resolution are not only expressed through the link between class level and aggregation level, but also by the spatial variability of the attribute values.

Object aggregation often requires different class hierarchies. In many cases object aggregation will imply a completely different thematic description of the objects, so that new classes should be defined. This is illustrated in Fig. 7.7 where farm yards and fields have been aggregated into farms and these in their turn into farm districts. The aggregation hierarchy has a bottom up character in the sense that starting from the elementary objects composite objects of increasing complexity are constructed in an upward direction (in Fig. 7.7 from left to right). The farm districts should only consist of farms and farms should be mutually adjacent so that the adjacency graph (see section 7.4) of the farms belonging to one district is connected.

The aggregation steps in Fig. 7.7 show how fields are considered as elementary objects at level 1. They are defined per growing season as spatial units under one crop.

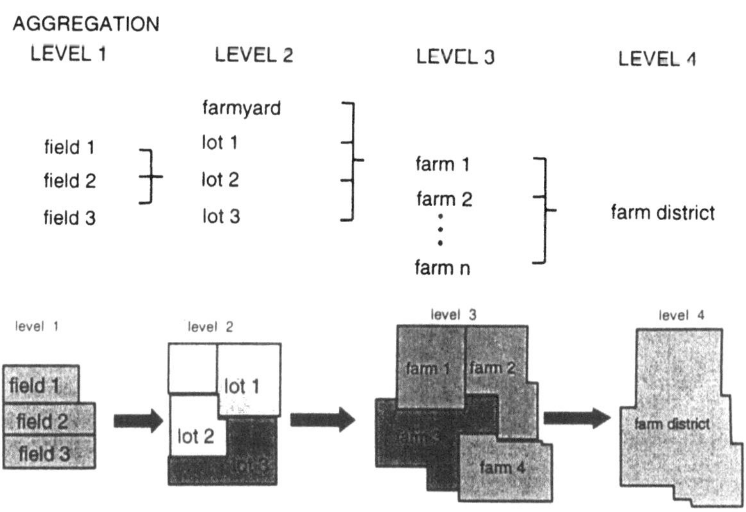

Figure 7.7 An example of a functional object generalization process.

For the farmer they are management units, because management operations are planned and performed per field. They are aggregated to lots which are elementary objects at level 2, i.e. these objects belong to the extensions of classes such as "arable-lot" and "grass land". These are management units at a higher level; the farmer will maintain a drainage system per lot and he will decide per growing season how to partition each lot into fields. These lots might both belong to a superclass "farm lots" in a land use data base and these again might belong to an even higher superclass "lot" which also contains the classes "forest lot" and "residence lot". The aggregation step from level 1 to 2 and the next steps to the levels 3 and 4 where we have the farms and farm districts show that after each step new objects are created. At farm level the farmer will decide whether he will be a cattle farmer or whether he will grow arable crops, in the latter case he has to decide on a rotation scheme. At district level the infrastructure and irrigation schemes will be developed. The objects at each level have their own thematic description expressed in an attribute structure that should be defined in a class hierarchy according to section 7.2. In this example each aggregation level requires its own classification hierarchy. This is structured so that the generated attribute structures provide the information to support the management operations defined at the aggregation levels of the objects. Fig. 7.8 represents a classification hierarchy for each aggregation level.

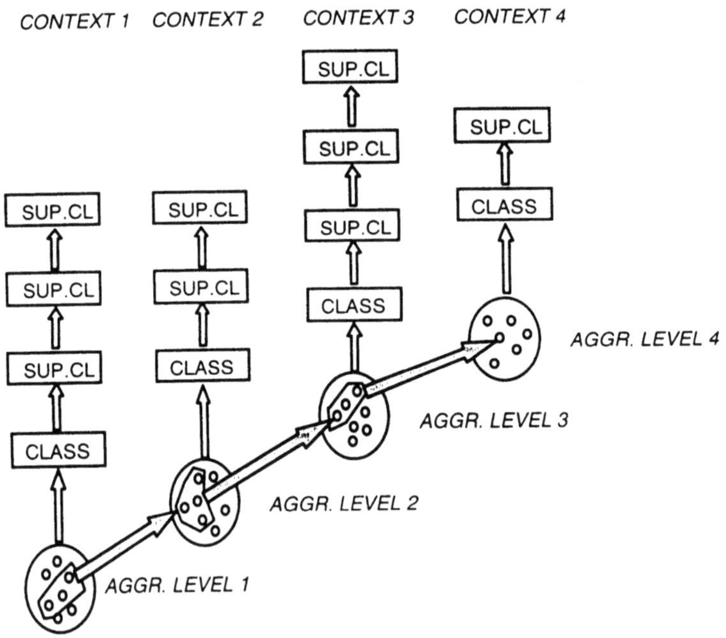

Fig. 7.8. Classification hierarchies related to aggregation levels.

This is an example of a more general situation where objects at each aggregation level are functional units with respect to farm management. We could take as well examples on ecological development, or demography and many more. Each aggregation level within such a hierarchy has its own (sub) context within a thematic field, expressed through a class hierarchy with related attribute structures. The different (sub) contexts are related by the fact that sets of objects at one level can be aggregated to form the objects at the next higher level. There are often also relationships between various classes of the different class hierarchies related to the aggregation levels as was the case for the cover classes for the farm lots, the farm types and land use types of the farm districts at the levels 2, 3 and 4 of Fig. 7.7.

There are bottom-up relationships between the objects at different levels in the sense that the state information of the lowest level objects, as contained in the attribute data, can be transferred through a process like Fig. 7.6, to give state information about the objects at higher levels. There are top-down relationships in the sense that the behavior of lower level objects will be constrained by the information contained in the higher level objects.

7.3.4 Structural object generalization

Structural object generalization will be explained by means of an example based on a database where the spatial description at a 1:50.000 scale, of a drainage system. The database has been structured according the FDS as in section 7.3 (Martinez Casasnovas 1994). A generalization process is executed to derive the 1:100.000 representation, so that we reduce complexity to stress spatial structure referring to the network structure of the drainage system in relation to the subcatchments. Generalization will keep the area of the aggregated subcatchments and the network structure of the system invariant. Therefore, computation of overland water flows per node in the network will not be significantly effected.

The database contains geometric data and thematic data of the elements of the system, as defined in the example of 7.2.1. Let the attribute `ORDER` contain the Strahler order number of each drainage element. These numbers in combination with the function `Upstr[ ]` allow to analyze the stream network built by the drainage elements. Through this network aggregation steps can be defined for the catchment areas, following Richardson (1993, 1994).

Structural object generalization starts with identification of drainage elements that are not mappable at the target scale, those are the elements with Strahler number = 1 with an average width (aw) < `Thr(eshold)`. The minimum mapping width of the drainage elements will be put at 0.75mm, that gives a threshold `Thr=0.75` mm·scale^{-1} at terrain scale, hence `Thr=75` m in this case. For the drainage element D_i, aw_i can be computed from its $AREA_i$ divided by the $LENGTH_i$ of its water line W_i:

$$aw_i = AREA_i / LENGTH_i$$

The selection procedure applied to the drainage elements is then

- select the drainage elements D_h with $ORDER_h>1$
- select from the class with $D_h = 1$ the elements D_i with $aw_i \geq Thr$

The set of elements that should be eliminated is then

$$S = \{D_i | ORDER_i=1, aw_i<Thr\},$$

their catchments should be combined with adjacent catchments to form aggregates. Elimination of the drainage elements $D_i \in S$ should consist of the following steps

eliminate W_i $AGGR(D_i, C_i) = C_{DHi}$ find C_h for which $Upstr[C_h, C_i] = 1$ $AGGR(C_{DHi}, C_h) = C_{h;i}$

where the notation C_{DHi} means that the area of D_i has been merged into the area of its subcatchment, the notation $C_{h;i}$ means that the area of C_{DHi} has been merged into the area of C_i. When water line W_i joins the outlet point $End(W_i)$ with only one water line W_j of a drainage element that will not be eliminated then the catchment of W_j should be merged with $C_{h;i}$

find C_j for which $Upstr[C_h, C_j] = 1$ if $D_j \notin S$ and there is no $D_{k,j} \notin S$ with $Upstr[C_h, C_k] = 1$ then $AGGR(D_j, D_h) = D_{jh}$ $AGGR(C_j, C_{h;i}) = C_{jh;i}$

The notation D_{jh} means that D_j and D_h have been merged and $C_{jh;i}$ means that C_j and $C_{h;i}$ have been merged. When these steps have been done for each element $D_i \in S$, then new Strahler numbers can be assigned to the remaining elements according to their new position in the network. Selection can be repeated until no more elements are eliminated.

A Test Case

The drainage system represented in Fig. 7.9a is used as an example. It is a schematic representation of the Romani Drainage system in the Anoia-Penedes Area in NE-Spain (Martinez Casasnovas 1994). The total area of this drainage system is 28.53 km^2, at a 1:50.000 scale it has 37 mappable drainage elements. The figure only shows the water lines W_i with their catchments. Drainage elements D_i are not shown.

Transition will be done from the original scale to the scale 1:100.000. First drainage elements with Strahler number 1 and width < 75m are eliminated, their catchments are aggregated with their downstream catchments. Then the remaining drainage elements are reclassified and the procedure is repeated until no more elements are eliminated (Fig. 7.9b). The drainage system represented at the 1:100.000 scale has

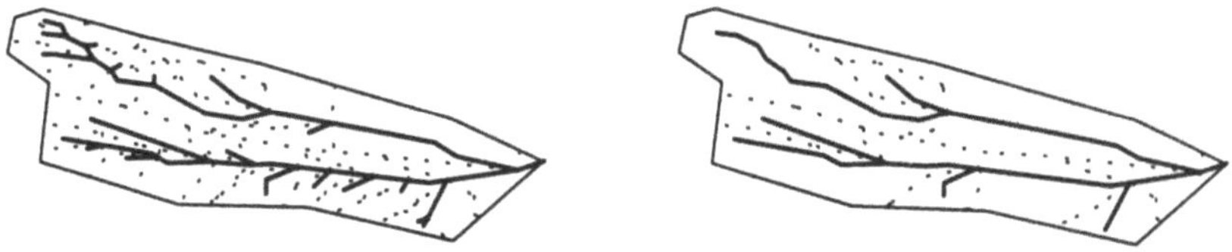

Figure 7.9 a) The drainage system at 1:50.000 scale representation
b) The drainage system at 1:100.000 scale representation

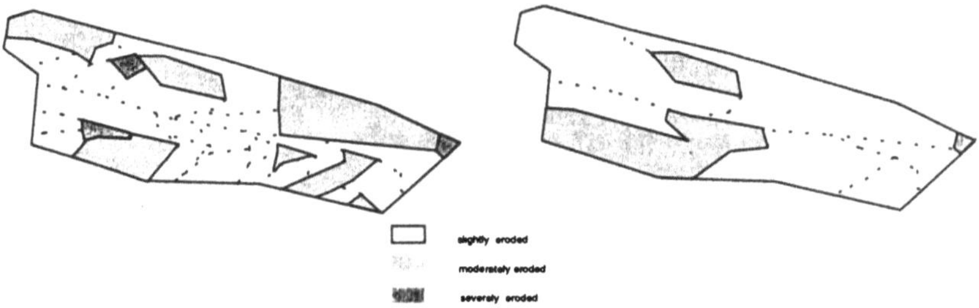

Figure 7.10 a) Erosion classes estimated in 1:50.000 scale representation
b) Erosion classes estimated in 1:100.000 scale representation

only nine mappable drainage elements. Their catchments are aggregates of the catchments shown at the 1:50.000 scale. This implies that the information carried by the original catchments is now transferred to these aggregates.

The Romani drainage system has been mapped for an erosion survey. Erosion classes are estimated per catchment from information contained in attributes of the drainage elements. These are used to compute per catchment the drainage density [km·km^{-2}] and the crenellation ratio [km·km^{-2}]. These data combined with depth and activity class of drainage elements determine the erosion class of each catchment. When the area of the catchments are summed for each erosion class we find in the original situation at 1:50.000 scale that 68% of the area is slightly eroded, 30% is moderately eroded and about 2% shows severe erosion (Fig. 7.10a).

After generalization, aggregated catchments are formed for which erosion classes have to be estimated. Although a large number of drainage elements have not been represented any more at the reduced scale, the information they carry has been transferred to the aggregated catchments. With these data we find now for the erosion classes that 79.5% of the area is slightly eroded, 20% is moderately eroded and 0.5% is severely eroded (Fig. 7.10b). These numbers deviate from the original values. Furthermore, the spatial distribution of the occurrence of the erosion classes differs from the original distribution. This would have been even worse if we had completely ignored the information carried by the eliminated drainage elements. Then the values would be 99.5%, 0% and 0.5%, respectively.

The structural generalization of the drainage system kept its constituting entities as hydrologic units. This had the effect that computation of hydrologic processes is invariant after generalization. The generalized network could, however, not be used to formulate reliable statements about erosion classes of the areas in the system. That would require another generalization where we had to specify what statements about erosion should be invariant after transformation. A class driven or a geometry driven strategy might have been more useful in this case.

7.4 Object generalization and levels of spatial complexity

In section 7.3 we discussed several strategies for generalization of spatial databases. These strategies were based on the concept of spatial object aggregation in combination with class hierarchies. For object aggregation the information of lower level objects is aggregated to higher level objects, but in principle the original detailed information is maintained. It remains possible to access the detailed information of the lower level objects through the aggregated objects. This leads to a less detailed terrain description, that may be compared to the result of a map generalization.

Such output could be used as input for a following aggregation step. This has been illustrated in Fig. 7.11, showing aggregations starting from the situation in Fig. 7.4c. Regions in that figure are assigned to more general classes in Fig. 7.4d and aggregated to form the larger regions in Fig. 7.4e.

Class generalization and object aggregation of Figs. 7.4 and 7.11 have been represented again in a different way in Fig. 7.13, which combines per database generalization step two steps like those of Fig. 7.4. In the first step the objects of the different classes are assigned to the super classes at the next higher level in the hierarchy (compare the class generalization step of Fig 7.4). In the same step objects forming a

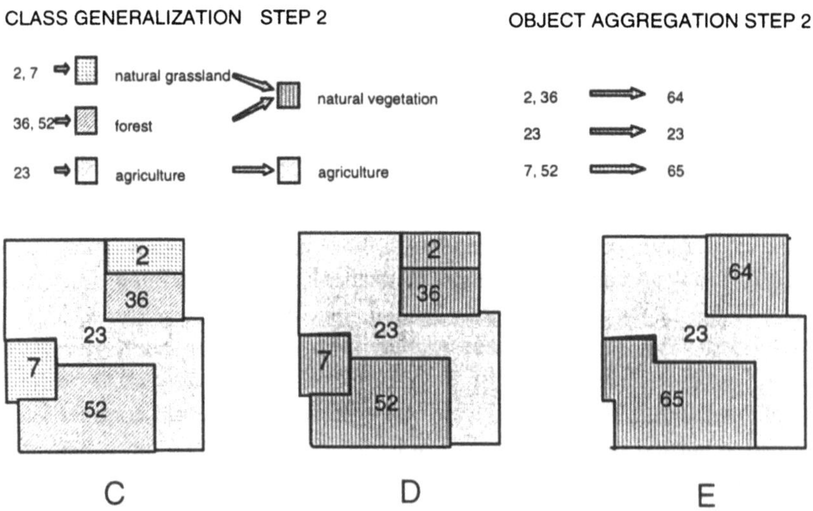

Figure 7.11: The second aggregation step for the objects of Figure 7.4c.

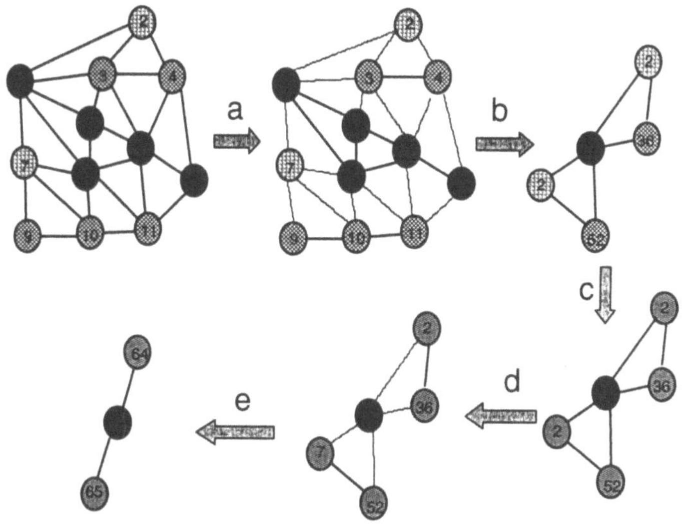

Figure 7.12: The adjacency graphs related to different stages of the generalization processes of Figures 7.4 and 7.11.

region in a super class are aggregated to form a larger object (compare the object aggregation step of Fig. 7.4). This procedure is repeated in the second step of Fig. 7.13.

This figure shows that the two steps of the example of section 7.3.1 reduce the number of objects, that is why we prefer to talk of database generalization because the process generated objects with a lower spatial and thematic resolution than the original objects. Because the original objects formed a geometric partition of the mapped area and because generalization used the topologic and hierarchical structures in which the objects had been modeled, this resulted in a new set of objects that also formed a geometric partition of the mapped space. It yielded though a terrain description of a reduced spatial complexity as is shown in the stepwise reduction of the complexity of the adjacency graphs of Fig. 7.12.

Each object is represented by a node in these graphs and the adjacency between two objects is represented by an arc. This figure gives the adjacency graphs related to each stage of a process that starts from the situation B of Fig. 7.3 where the original objects have been assigned to their super classes. If we follow the steps of Fig. 7.12 then we see that:

- in step a the regions per class have been identified,
- in step b the objects in each region are aggregated to form a composite object that is represented by one node,

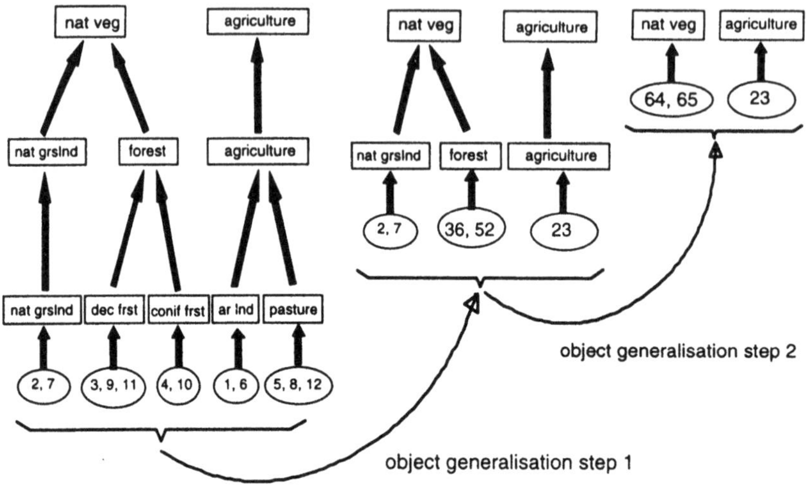

Figure 7.13 A diagram representing the object generalization steps of Figures 7.4 and 7.11.

- these regions are after step c assigned to more general classes,
- in step d regions at this higher class level have been identified, these are composed of the objects obtained after step b,
- then finally after step e each of these regions have been aggregated again to form the objects at the higher aggregation level which is then represented by one node, this is the adjacency graph of Fig. 7.11e.

Reduction of spatial complexity is one of the important aspects of generalization, traditionally applied to reduce the information content of a map so that a mapped area could be represented at a smaller map-scale. It has two steps: conceptual generalization and graphic generalization. Conceptual generalization results in redefinition of mapped spatial features or objects to reduce their number for terrain description at the smaller scale. Graphical generalization is a simplification of the graphical representation of these features or objects, including aspects as geometric simplification, object displacement, resymbolization etc.

7.5 Conclusion

When we deal with spatial database generalization in a GIS the main aim is to achieve a simplified terrain description, i.e. a lower spatial complexity to emphasize spatial patterns and relationships that might be difficult to find in a more detailed terrain description. This might include the graphical representation as well, but that is not necessarily so. It means that spatial database generalization is highly related to conceptual generalization, its main aim being data reduction. We have seen that to a large extend it can be defined as operations for databases that are implementations of the formal data structure as explained in section 7.2.

A spatial database may contain information about different aspects of a particular area, as we saw in the example of section 7.3.4. Generalization may keep one aspect invariant, say the primary interpretation of the database. The other aspects may be affected so that their information is not reliable after the process, say the secondary interpretations of the database. Generalization as a type of transformation of a spatial data base requires explicit decisions about which aspects of the original data bases to keep invariant, and hence to what to be considered as the primary interpretation of the data base. This choice will be made within the users context of the data base, i.e. the user will be interested in the correct representation of some spatial characteristics, while others may be deformed by the transformation.

A good understanding of database generalization may be useful for the design of procedures for spatial data acquisition. Information extraction from images is partly a reverse process to generalization. Generalization is a process with a stepwise data reduction, going from high resolution to low resolution. Information of the high resolution objects is merged into low resolution objects. Image interpretation can often be formulated as a process where data are produced stepwise. We can learn from generalization processes what information low resolution objects carry about their constituting high resolution objects. This knowledge may help us in image interpretation, where large image segments can be seen as low resolution objects. These should then contain thematic information in addition to the radiometric and spectral information of the image itself, to identify smaller segments that may represent high resolution objects.

Exercises

1. Consider implementation of the pea model from chapter 1 and suppose that sufficient data are available to implement this model in a GIS. Suppose that a very detailed soil map is available. Develop some general procedures for the implementation.
2. Consider the various requirements for data as expressed in chapter 6. Discuss how these requirements could interfere with linking and models as presented in this chapter.
3. There are various ways expressed in chapter 5 on upscaling, whereas the current chapter deals with generalization. Discuss advantages and limitations of the two approaches for a practical case study.
4. Take the solar radiation model from chapter 3. Describe the different forms of generalization and data requirements when such a model is to be applied at the field scale, the farm scale, the regional scale and the national scale.

IV Perspectives for automated modelling

Modeling has been so far primarily been treated as a highly skilful activity and as a handicraft. Computers and automation were necessary to speed up calculations, to visualize inputs and outputs, and to efficiently combine layers of information. At this stage, however, we can distinguish some new trends, which could be beneficial in the near future. These trends are based on increased facilities for automated calculations and developments. This part addresses some of these elements and has therefore a more exploratory and reflective character than the previous parts. We focus on object oriented programming and on trends in automated modelling.

8 Examples of object-oriented design in agricultural research

PJ Schotman

8.1 Introduction

Object-oriented techniques have become widespread in software engineering. One of the newest developments in object-oriented technology is the specification and use of design patterns. This paper shows how two of these design patterns can be used in (agricultural) research. The "Template Method" design pattern applied to modeling and simulation shows the benefits of object-oriented technology with respect to model adaptation. The use of this template together with a proper class hierarchy could be a crucial step in the development of a reusable model base. The "Composite" design pattern is shown as a second example and can be worthwhile in GIS applications.

The rise of object-orientation reflects and recapitulates the history of computing as a whole. The earliest works in computing concerned themselves with what we now call 'programming'. Only later did a conscious concern with design and analysis as separate issues arise. Similarly, it is object-oriented programming that first attracted attention, later design and analysis issues became dominant (Graham 1994).

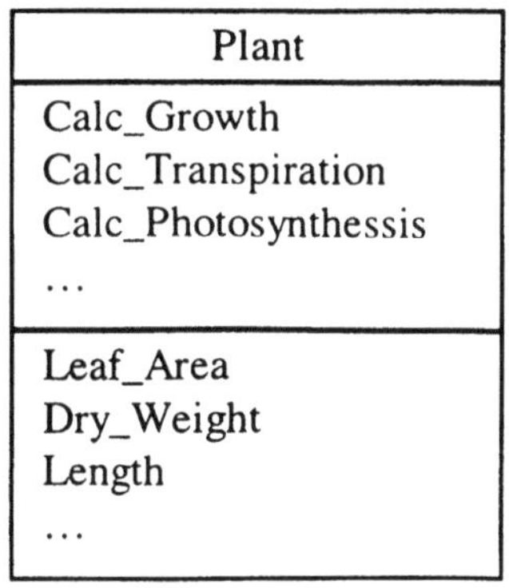

Figure 8.1 The class of plants with some methods and attributes

Object-oriented programming languages (OOPL) have extended functionality as compared to their conventional counterparts. In some languages this extension is simply added to the already existing non-object-oriented language specification such as C^{++} (Stroustrup 1991), whereas other languages such as Java are defined around the concept of object-orientation. Language implementations also differ extensively, in some implementations every language element is an object (e.g. Java or Smalltalk) while in others (e.g. C^{++} or Fortran90) this is not the case.

Object-oriented programming (OOP) together with its pendants object-oriented design (OOD) and object-oriented analysis (OOA) could be seen as way of thinking when solving problems that require programming. Booch (1994) defines an object as follows: "An object has state, behavior and identity; the structure and behavior of similar objects are defined in their common class". Fig. 8.1 shows an example of a class; objects (or instances) are unique instantations of classes in that they combine the structure (attributes or slots) and behavior (procedures or methods) that is defined for the class. Structure and behavior are defined at the class level so that they can be used at the object level.

8.2 Object-oriented technology

8.2.1 Promises of object-oriented technology

Graham (1994) uses 15 pages to elaborate on the advantages of object-oriented methods, the following are highlighted in this paper:

- Easier modeling of the real world. Modeling a problem in terms of components that have both structure and behavior seems to be very natural. The meaning and semantics of real world objects can be captured more effectively. Modeling real world objects into hierarchies and reasoning on the basis of these hierarchies has been a prominent research subject in the field of artificial intelligence (Rich and Knight 1991).
- Productivity improvement. Reuse of earlier software engineering efforts is considered a predominant advantage of object-oriented method. However, this advantage has proven difficult to realize (Pancake 1995; Meyer 1995). User interface builders are a classical example of code reuse. Reuse use to be mainly centered upon code reuse, however, it is only until recently that it was realized that design reuse may also be of practical value (Schmidt 1995).

In agricultural research, modeling and simulation is of increasing importance. Researchers are building simulation models themselves and/or use (and usually modify) models built by others. There are advantages when adopting an object-oriented approach, the most prominent may be the easy insight in the structure and workings of a model and the way parts can easily be modified to suite ones needs. It is not surprising that the first object-oriented programming language (SIMULA67) was specifically intended for simulation (Dahl et al 1982). Unfortunately, owing to insufficient commercial support this language has not become a success.

8.2.2 Basic features of object-oriented technology

Apart from the already mentioned coupling between structure and behavior, object-oriented technology is generally centered around the following three attributes: inheritance, polymorphism and encapsulation. The latter two can be considered as special cases of abstraction (Graham 1994). Here they are considered separate.

Polymorphism

Polymorphism applies to the behavior of an object, it refers to the way objects of the same class respond to a message that is sent to them. Messages are calls to procedures that belong to the interface of an object. Objects communicate through these calls. Messages can be considered as imperative statements that are to be executed by the receiver (e.g. `plant_47.length` corresponds to "plant_47 return your length"). Messages to different objects may share the same name although their implementation may be different. Consequently, upon execution, the resulting chain of events may also be different.

Inheritance

Inheritance refers to the process of obtaining attributes and behavior among classes. Given the `Plant` class in Fig. 8.1 a sub-class `CAM_Plant` could be defined which would inherit all of the attributes and most of the behavior of the regular `Plant` class however, some behavior will be different. The class `CAM_Plant` may also have additional attributes. In this particular case, the photosynthesis process (amongst other things) will be different, therefore, the method that describes the photosynthesis process will have to be different. The standard method for `Calc_Photosynthesis` defined for the class `Plant` will be overridden with a new method (with the same name) belonging to the class `CAM_Plant`. All objects that are instances of the `CAM_Plant` class will apply this new method and behave differently with respect to this procedure compared to objects that are instances of the `Plant` class. Note that for behavior that is not overridden instances of both classes will behave exactly the same. Object-oriented programming languages have (internal) logics to determine the most applicable method for a given object.

Inheritance relations result in classes hierarchies, sub-classes relate to their parent classes through the `IS-A` relation. Other hierarchies are also possible, for instance, the `PART-OF` hierarchy. This hierarchy relates objects to one another and is prominent in geographical information systems (GIS). An object-oriented design for this relation will be given later.

Encapsulation

Encapsulation refers to the interface of objects. The interface of objects is defined at the class level and it contains all messages to which objects of that class can respond. Through encapsulation it is possible to hide implementation details from the object's use. By doing so a software engineer creates a certain amount of freedom. This freedom allows him to easily change details of certain classes without disturbing their use. Some programming languages (e.g. C^{++}) offer specific identifiers to mark certain methods and attributes as hidden, others do not (e.g. CLOS (Keene 1989)). Some simple software engineering directives, however, (e.g. "Never access an attribute of an object directly.") also suffice to hide the implementation details of a class.

8.2.3 Support for object-oriented software construction

There currently consists a wide range of methods (or methodologies) and tools that help the software engineer both during the analysis of a problem and design and construction of the software solution. The most prominent methodologies seems to be: OODA (Booch 1994), OMT (Rumbaugh et al 1991), OOA/OOD (Coad and Yourdon 1991a; 1991b) and OOSE (Jacobson et al 1992). Van Bergeijk et al (1995) show the use of the OOSA methodology (Shlear and Mellor 1992) in the agricultural domain. Tools that support these methodologies range from informal to formal. Tools that support formal design specification methods sometimes allow for automatic (complete) source code generation or template generation (i.e. name, class and interface of methods are specified but the actual workings have to be filled in).

Above and other methodologies have been extensively compared and reviewed, for instance in: Wirfs-Brock and Johnson (1990), Monarchi and Puhr (1992) and Wilkie (1993). They will not be discussed further.

8.3 Design patterns for reusable object-oriented software

Reuse of earlier software engineering results has proven difficult to realize for many reasons (Meyer 1988). Design patterns may be promising technique for widespread reuse of software architectures (Schmidt 1995). Design patterns stem from architecture, Christopher Alexander (in Gamma et al 1994) describes a pattern as follows: "Each pattern describes a problem which occurs over and over again in our environment, and then describes the core of a solution to that problem in such a way that you can use this solution a million time over, without ever doing it the same way twice.". Design patterns explicitly capture knowledge that experienced developers already understand implicitly. Using this knowledge can lead to better designed software because it contains the craftsmanship of experts in the field. Design patterns also function as a vehicle for communication between the developers and users of the source code. In the next sub-sections two templates from Gamma et al (1994) are taken as an example for problems that are important in research related to production ecology (PE).

8.3.1 Simulation as an example of the "Template Method" design pattern

Reuse and alteration of simulation models built by others frequently occurs in PE research. In many cases, careful inspection of the source code listings (usually FORTRAN) is necessary to determine the place of one's own changes. Problems become serious when these changes imply changes to the interface of one or more sub-routines in the model. After some time the model may have evolved significantly and, unless the changes have been documented carefully, differences between this adapted model and the original one may be difficult to determine. Applying the Template Method design pattern can help facilitate this well known problem. Changes easily stand out and different versions become apparent through the class structure.

The Template Method design pattern can be used to define the skeleton of an algorithm of an operation deferring some steps to subclasses. The pattern lets subclasses redefine steps (called: primitive operations) of the algorithm without changing its structure (Gamma et al 1994). As an example, the calculations per time step in a simulation model are being viewed as a template method.

Fig. 8.2 shows, in OMT notation, a simple class hierarchy for which the template method `Calc_Photosynthesis` is shown. This template is defined for the abstract class `Crop`. For brevity, the figure does not show the attributes of the class. The arrows indicate inheritance relations (i.e. `IS-A` relations). Abstract classes and abstract primitive operations are shown in italics. Abstract primitive operations (also called "hook" operations) do nothing in the abstract class and are meant to be overridden. Abstract classes should not, as a matter of design principle, be used directly. One should first define a concrete sub-class, then define the behavior of the not yet defined hook operations, hereafter the concrete class and its interface are ready for use.

An important goal in designing template methods is to minimize the number of primitive operations that the final concrete sub-class needs to override. With the use of predefined models (i.e. template methods) in a model base this should certainly be the case, otherwise the model base will be of limited value.

The primitive operations that constitute the template method `Calc_Photosynthesis` are:

```
Get_measured_Light
Calc_Light_above_Crop
Calc_Light_Distribution_in_Crop
Calc_Photosynthetic_Rate
```

The actual implementations of two these primitive operations are also defined for the `Crop` class. As can be seen in the figure, the abstract classes `Greenhouse Crop` and `Field Crop` define different methods for `Calc_Light_above_Crop`. Obviously, the method for `Field Crop` is rather simple, it just passes on the measured light, while the method for `Greenhouse Crop` can be much more complicated because it must take the greenhouse construction into account. The actual greenhouse will be stored as an attribute in one of the slots of the actual crop object. In an object-oriented implementation this greenhouse will most likely be an object that has specific information concerning its roof structure and a method to calculate its (dynamic) transmissivity. In Fig. 8.2, sub-classes of `Greenhouse Crop` define methods for `Calc_Light_Distribution_in_Crop`. When a researcher devises another planting arrangement he or she should create a new sub-class of `Greenhouse Crop` and redefine the method for `Calc_Light_Distribution_in_Crop` for this sub-class.

The use of the `Calc_Photosynthesis` template method is simple. The use of an object is only possible when it has been created and initialized (i.e. appropriate slots such as `LAI`, `Dry_Weigth` are assigned an initial value). Now the message `greenhouse_crop_23.Calc_Photosynthesis` is sent by some client. Typically, this message is being sent during a simulation run.

Some additional remarks:

- Template methods can be nested. The `Calc_Photosynthesis` template method could be part of the `Calc_Growth` template method.
- Elaborate and simple models of the same process can live along side of each other in parallel class hierarchies. Differences in template methods are distinguished through the class structure. For instance, under the class `Crop` two subclasses could be defined `Elaborate Crop` and `Simple Crop` each having a template method for `Calc_Photosynthesis`. The sub-class structure below both classes can be the same, however it is likely that the `Elaborate Crop` contains a more detailed sub-class hierarchy. Since research is a dynamic activity, results of

research, for instance, model bases grow and change over time, template methods can/will also change over time. New versions of a model base will contain improved template methods and/or improved primitive operations. Version control procedures for software can be used as a method to document those changes.

- It is possible to opt for different strategies with respect to structural and temporal changes of the class hierarchy. *Temporal* changes carried out by *users* of the hierarchy first involves creating a sub-class below the *lowest* level in the hierarchy (in Fig. 8.2 this would mean either below `Full Field Crop` or `Row Crop`). Second, redefinition (overriding) of the method that the user wants to change is carried out. This could be any method of the interface (i.e. all the methods inherited) of the newly defined class. Permanent changes will be carried out by the maintainers or designers of the model base and involves restructuring of the

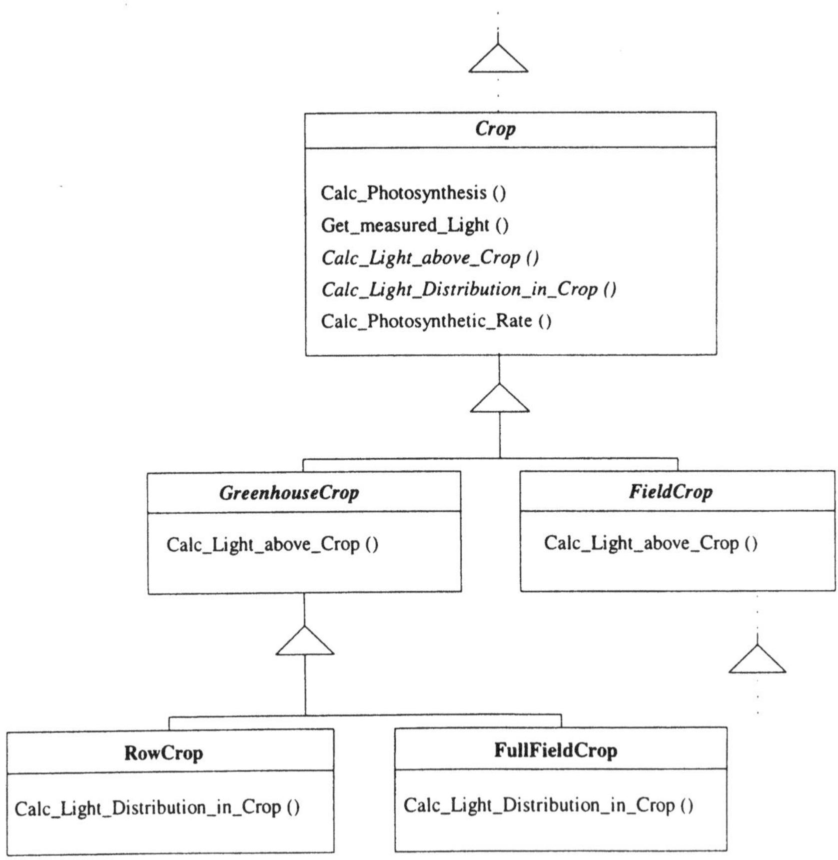

Figure 8.2 A simple class hierarchy in which classes with methods are shown

hierarchy, that is, inserting new classes at higher levels and implementing appropriate methods for these classes. Temporal changes can become permanent (thus inserted at the appropriate level) when they serve a more general use.

Data belonging to an object can be stored in the attributes or slots (data members in C++). This can make the list of arguments of methods simple because the attributes of a class are available and need not be supplied as arguments when defining a method. In an object-oriented design forcing function input (e.g. measurements of the environment) may be retrieved by sending a message to an environment object that carries this data. For example, `greenhouse_compartment_23.Temperature` returns the current temperature.

To summarize, the Template Method design pattern shows a clever use of polymorphism. Polymorphism is important in model bases.

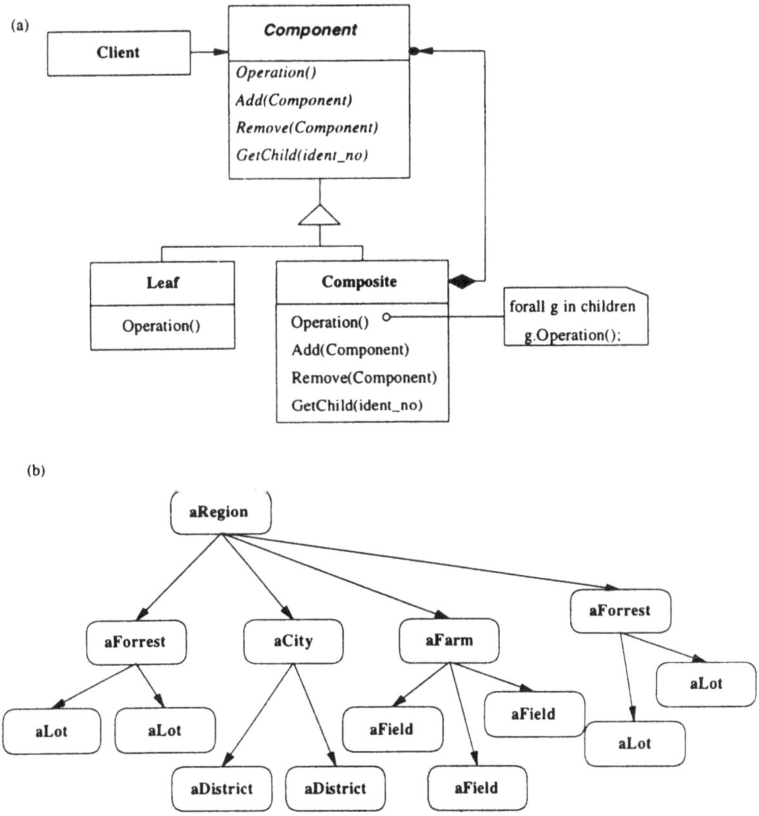

Figure 8.3 The Composite design pattern (a) and an example (b)

8.3.2 The "Composite" design pattern as a basis for scaling

The Composite design pattern has been intended to compose objects into tree structures. These tree structures represent part-whole hierarchies (through the PART-OF relation). The basic idea behind this pattern is the uniform treatment of individual objects (or leaf nodes in the tree hierarchy) and composites (i.e. sub-trees). The Composite pattern can be found in graphics applications like object-oriented drawing editors and almost every interface toolkit (Krasner and Pope 1988) available today. In research, composite structures are typical for GIS applications.

Fig. 8.3a shows the structure of the design pattern in OMT notation. In a geographical information system, the abstract class `Component` could be called a GIS Object. All kind of abstract operations could be defined for this abstract component. These operations should be filled in for the appropriate composite and leaf classes. It should be noted that there can be many concrete `Leaf` and `Composite` classes. They may also be arranged into class hierarchies themselves. For instance, the `Lot` and `Field` class will have many attributes in common, they could both be sub-classes of the abstract class `Rural Unit`. The Composite design pattern allows for an easy interface for a "client" that calls a function on (i.e. sends a message to) an object in a specific part-whole hierarchy (like the one in fig 8.3b). Fig. 8.3b shows a simple object hierarchy, the arrows indicate the PART-OF relation among objects. An object hierarchy like the one in Fig. 8.3b, it can be made as detailed as an application requires. There can be many objects of the same type all representing different individual entities.

With the easy interface for clients, there are some additional responsibilities to attend to. For instance, some of the operations defined for the `Component` only apply to composites, appropriate error trapping needs to be defined for primitives (e.g. it is not allowed to add a component to a leaf object). Also, one would like the logic of an application to be automatically enforced which means that one does not want every composite to be part of another composite (e.g. in Fig. 8.3b: `aCity` cannot be part of `aForrest`).

In the simplest case, operations defined for composites could be an accumulation of conceptually the same operation defined for its primitives. However, this does not need to be the case and at least provisions should be made to integrate the outcome of the operation for different leaf classes. An operation like `Calc_Production` called on `aRegion` in Fig. 8.3b will result in a combination of agricultural products (all `aFarms`), an amount of lumber (all `aForrests`) and the production of `aCity` (which will be the combination of industries present in the city). One could also define `Calc_Production` for `Regions` to return something like the economic gross product of the region. Possibilities are infinite, the design pattern supports scaling in a very natural way.

8.4 Conclusions

Two design patterns have been shown to illustrate the possibilities of object-oriented technology in the domain of production ecology research. The main advantage of the ideas presented in the examples above will be its reuse by people other than the developers. This immediately shows the drawback: object-oriented model bases are not yet available and it will require some effort (i.e. money) to develop them. However, this investment will easily be recovered when researchers are going to use and extend the model base. Within the GIS community the situation may be better, because object-oriented geographical information systems have already been reported (Burrough 1996).

Apart from the advantages in the more traditional ways of modeling, newer modeling approaches like individual modeling (e.g. Hogeweg and Hesper 1990) or multi-agent modeling (e.g. Attonaty and Pasquier 1996) will profit even more of an object-oriented implementation.

Exercises

1. Develop some statements to make an object-oriented program for the pea model from chapter 1.
2. Sketch how object-oriented programming could function in a GIS. Consider elements like upscaling, accessing multiple layers of information, displaying. Take, if of interest, a simple model (the global radiation model) and an area of the size of a region.
3. Object oriented programming is of particular interest as long as we consider the space of interest to be composed of individual objects. This could function well in some settings (like those described in chapters 1 and 7), but in continuous spaces as mentioned in chapter 6 this could fall short. Here the continuous approach appears to be more useful. Discuss these shortcomings, and try to formulate extensions or alternatives.
4. - Develop a simple object-oriented model of a prototypical arable farm. Consider that the farm grows various crops. Make two class diagrams (hierarchies): one for the physical/geographical structure of the farm and one for the crops. (Hint: use the composite design pattern for the physical/geographical structure of the farm.);
   - Add some attributes (data members) and methods (member functions) to both diagrams, consider for instance where to place methods like: "calculate yield".
   - Suppose your objective is to calculate the farm's yield with various product combinations at different locations across the globe. Add sufficient detail such that the yearly yield (in financial units) can be calculated;
   - Optional: program the lot in your favourite object-oriented programming language.
   - Suppose you want to extend the model such that it can also be used to calculate the yield of mixed farms. Is it necessary to add another class hierarchy for farm animals or are there other solutions possible? And how may the hierarchy that represents the physical structure of the farm be extended?

9 Computer based trends in automated modeling

JL Top

9.1 Introduction

The art of computer-based modeling is changing rapidly. In the near future, automated modeling will exploit information technology to turn the art of modeling into a profession. In this chapter we review a number of advances to be expected in the near future. Two important developments are:

- explicit formalization of background knowledge and
- structuring of the modeling process.

Explicit knowledge will be represented as mathematical and logical descriptions. Yet more important, conceptual languages will be developed to link these often abstract models to a common sense interpretation. A model, however, is always concrete as an answer to a question. Support for asking the correct question, thus guiding the modeler to an effective, high-quality answer will be necessary.

Automated modeling will exploit information technology to turn the art of modeling into a profession

Modeling is a basic human function that pervades all our conscious activities (Rothenberg 1989). For example, even a simple statement saying: 'It's cold outside' already makes up a model. It tells us something about a part of the world without immediate observation. Viewing this statement as a model, it reveals some important aspects of modeling:

- *Generalization:* 'Cold' is a generalization that we can use without referring to a concrete object. A model is based on abstraction, allowing to share information without giving a complete description of a system. Completeness is impossible. Moreover, generalization not only occurs across the objects to be modeled, but also across modelers. In other words, modeling implies the need for a language to describe the world so that it can be understood by others.
- *Independent existence*: abstraction and generalization only exist in our minds. The model itself exists as a concrete entity. It has its own structure and properties. In this example it is a sentence in English with a certain intonation, loudness etc. The specificity of a model limits the range of what can be expressed by it.
- *Goal directness:* The statement *'It's cold outside'* can express different intentions: a warning, an explanation or simply an introduction. A model always exists within the context of a problem, a purpose. Therefore, the context of a model is necessary to understand it properly.

These features are also essential for *computer-based modeling,* the subject of this paper. First, by our ability to abstract (in our minds) we are able to understand simulations in terms of the modeled system. It is not always clear, however, to see what has (not) been modeled by inspecting a computer code. Therefore, software engineers use mathematics and logic to set up models in a formal way to reduce ambiguity as to how the model should be interpreted and what its limitations are. This is important because of the second factor: computer models are not abstract themselves, but consist of hard- and software. They are flexible, powerful and safe, but at the same time they are limited in expressiveness and performance. They can never represent the real system perfectly (Zeigler 1976).

The third issue, *goal directness*, is equally important in automated modeling. It requires explicit definition of a model. Building a model must be guided, such that the objective is met with as little side effects as possible. This means that ideally it should answer the question posed without loss of time and without the need for additional resources. Understanding the modeling process will help software and knowledge engineers to push the limits of modeling further into this direction.

In this chapter we will look at some trends in automated modeling, given the tension between generalization and specification that is essential in modeling. First, in Sec. 9.2 we consider the model as a bridge between theory and application. How can the contents of a model, i.e. of such a small-scale theory, be formalized? In Sec. 9.3 the modeling process will be explicated as an essentially interactive process based on assembling generic pieces of knowledge. We will conclude by summarizing the most important developments in automated modeling to be expected in the near future. The ideas presented in this overview are based on previous work on modeling (Top 1993; Top and Akkermans 1994).

9.2 Formalization: from theory to application and vice versa

Formalization of models - before implementation as a computer program - will be a major trend in the years to come (Akkermans et al 1993). Formalization means that information contained by models is represented in an objective (or better: inter-subjective) way, independent of how it will be implemented. By using well-defined general terms, ambiguity can be reduced. Moreover, the computer code can be generated almost automatically from the formal model. In this section we will take a look at the contents of models and to future developments that we can expect with respect to formal representation of these contents.

9.2.1 Modeling as small scale theory building

A model can be viewed as a small scale theory (Rosenbluth and Wiener 1945) dealing with a concrete question, but still general. In a black-box approach (e.g. 'curve-fitting'), the internal structure of a model is irrelevant, if it properly reflects the observed behavior by means of its inputs and outputs. It is implicitly assumed that it also will work for those input and output values for which no measurements were made: if the model were only valid for the measured values it would be useless. A truly black box model, however, cannot guarantee the quality of such an extrapolation. A usual, implicit, solution is to assume that the behavior of a system does not change significantly in the ranges considered. There is no possibility, however, to check whether this assumption holds.

More support for the validity of a model (without more observations) can be obtained by employing theoretical knowledge in the form of *domain laws*. Then, the quality of a model is not only guaranteed by validation through observation, but also because those theories have proved their value over time. Theory-based validation is sometimes called *confidence building* (Top and Akkermans 1994). It can be expected that the search for domain invariants and laws ('energy' and its conservation being an extremely powerful example in classical physics) will continue, providing a basis for the small-scale theories that our models are. As for computer science we will see information systems gradually change into knowledge-based systems.

9.2.2 Mathematics and logic

The most general invariants employed in modeling are elements of logic and mathematics such as variables, equations, clauses, operators, etc. They allow formalization of a model before implementation in a computer program. In a mathematical model, knowledge and data can be manipulated automatically and new values and statements can be derived without reference to the model context. All 'real world' aspects have been removed and the model has been reduced to a complex, but very well defined object. Computers happen to be good in doing formalized operations as they can detect syntactical errors and perform logical inferences. Therefore, logic and mathematics will remain primary players with respect to formalization of knowledge. In particular the following developments can be expected here:

- *Number crunching*. Increasingly powerful numerical and analytical algorithms are being constructed. This will not only affect the performance of simulators, but also make them less critical to the form the mathematical model.
- *Hybrid systems*. Continuous and discrete approaches will be integrated. This will allow for example to switch automatically between (parts of) models given their validity constraints.
- *Formal verification and validation*. Automated theorem provers will improve, thus allowing extensive formalization and model checking - before generating the computer code.
- *Non-linear analysis*. Recent interest in chaotic systems has given an important impulse to non-linear systems analysis. This will significantly increase the scope of phenomena to be represented.

9.2.3 Conceptual languages

The problem (and point) of formalization in terms of logic and mathematics is that the meaning of the variables and expressions is completely removed. For example, if we are modeling an electrical circuit we may want to know the voltage across a resistor. Our mathematical model could be $z = x \cdot y$. Without explanation it is unclear what the symbols mean. If the above law had been written as $V = I \cdot R$, using standard notation, it would have been immediately clear for those who are familiar with the domain. However, what if there are more resistors in the circuit? The need for explanation seems obvious, but all too often the meaning of variables and expressions is left implicit.

A mathematical or logical model without explanation is useless. The assumptions about when and why it can be used are unknown. Thus, a purely mathematical model cannot be reused or shared. What is needed is a conceptual description to provide a link with common sense understanding of the world. However, this conceptual level must be formal as well, eliminating ambiguity of natural languages, but at the same time it is less general than the mathematical level. For example, the network description of electrical systems is a kind of conceptual language, provided that the symbols are defined as ideal standard processes (Top 1993). Different languages will have to be defined for different application areas, with links between them.

An important function of conceptual languages is to hide complexity. For most cases the mathematical model can be derived automatically from the conceptual model. It has even been shown that computational (numerical) schemes derived from proper conceptual models in engineering are more effective (Van Dijk and Breedveld 1991) than purely mathematical algorithms.

Relevant in this context is the work done in *Qualitative Reasoning* (QR), a sub-field of Artificial Intelligence (Weld and De Kleer 1990). Researchers in this field claim to have filled the gap between mathematics and common sense. They propose a method for solving physical and other problems by defining a simple 'common sense' calculus based on a small subset of possible values (e.g. 'negative', 'zero' and 'positive'). This approach is supposed to deal with incomplete information in a way that is similar to human reasoning. However, even a simple second order dynamic system leads to a multiple set of solutions, of which most are spurious. In our view this approach can never be successful because it essentially remains a mathematical approach. Human reasoning is based on conceptual thinking and can therefore not be simplified in this rigorous way.

With respect to the conceptual level we may expect the following developments:

- *Graphical conceptual languages.* More and better conceptual languages will become available. In the field of knowledge-based systems much energy is already put into the definition of an ontology for several areas (Gruber and Olson 1994). These languages will be graphical rather than textual. This allows compositionality and direct syntax checking. Moreover, a graphical representation can more easily be inspected by the modeler (Karnopp et al 1990; Paynter 1961).
- *Model libraries.* Composite structures in terms of the above conceptual languages can be stored as a whole, with the underlying conditions for their application (Breuker and Van de Velde 1994; Top et al 1995).
- *Language integration.* Due to the proliferation of conceptual languages it will be necessary to restructure them and to allow for true multi-disciplinarity. Specific languages will be constructed on top of more specific languages, and languages of related domains will be linked. This will require an intimate link with theoretical research in the associated sciences, since new *invariants* should be defined.

So far we have only considered what can be expected with respect to the contents of models. However, the process of building a model is equally important and will dramatically change in the future.

9.3 Asking the right question

In a naive view of what modeling entails, one starts with some observations in the real world and from there constructs a model. This view - which is still dominating - misses the point in two ways. First, usually some domain knowledge is used to start the modeling process. Second and more important: model construction is not a single-step process, but essentially iterative. Building a model is as important as the resulting model itself, since many decisions about the problem context are being made along the way. The modeling purpose is gradually refined, until an answer can be provided with some assurance. In short: modeling is a *learning* process.

In Fig. 9.1 we show a schematic view of the modeling process. We distinguish three basic steps: specification of assumptions, construction of the model and assessment of the model. It starts with a (possibly vague) intention in the head of the modeler. Moreover, usually an initial model is available in one way or another. In traditional modeling approaches the emphasis has been on the construction step. Here, the building bricks (the *generic components* in fig. 9.1) were software statements in Fortran, Pascal or any other computer language. Then the model was assessed by comparing it with the measured behavior and it was adjusted where necessary. At some point the model was assumed to be good enough.

This approach, however, is inadequate in several ways. First, no guarantee can be given that the model fulfils the requirements of the modeler because these requirements are left implicit, i.e. in the head of the modeler and no specification step is performed. As a result, there is no control of the model quality and the model cannot be used for a different case, or shared with other modelers. Second, there is usually no

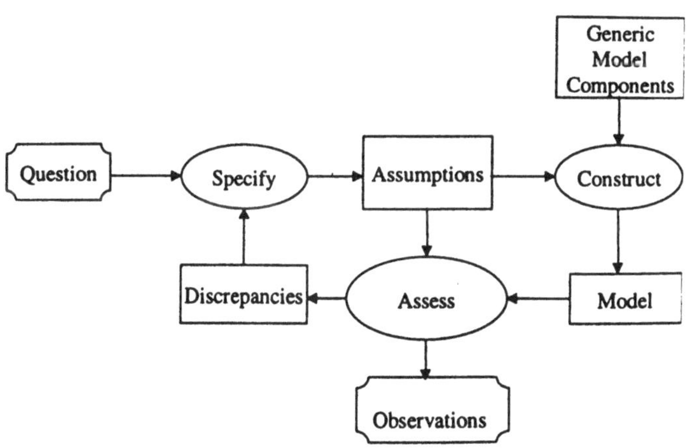

Figure 9.1 A general decomposition of the modeling task (From Top and Akkerman 1994).

distinction between different viewpoints (fig. 9.1). Computational and numerical considerations are easily mixed with conceptual issues. A typical example can be found in physical systems modeling: discrete changes in otherwise continuous model are often inserted as patches in the computer code. However, it has been shown (Strömberg et al 1993) that introducing an explicit switching concept at the conceptual level eliminates many problems both at the computational and the conceptual level. The complex computer code can even be generated automatically, and the modeler only deals with the much more appealing conceptual switch. The third reason is that no guidance is given on how to proceed. Specification of assumptions, assembling the model and assessing it against these assumptions is largely unstructured and is therefore only a matter of trial and error. Clearly, more structured approaches to modeling, based for example on the elementary scheme presented in Fig. 9.1, will be needed to help the modeler to answer modeling questions more efficiently.

With respect to computer support to modeling we can expect the following developments:

- *Assumption management.* Explicit bookkeeping of the modeling assumptions will be a major step forward in the availability of high-quality models. The background information *about* models that is nowadays given in textual form - or not at all - will be inherent to the models. This will help the modeler to ask the intended question.

Explicit bookkeeping of the modeling assumptions will be a major step forward in the availability of high-quality models.

- *Automated abstraction.* Based on general knowledge of what is typically relevant in a modeling context, some default operations can be done automatically, provided that the modeler is made aware of them. For example, in modeling dynamic systems it is possible to consider only processes that occur at a certain time scale. We may call this automated time scale abstraction.
- *Collaborative modeling.* By explicitly recording why and how a model is constructed, individual model fragments can be shared between modelers. Thus, modeling can become a collective rather than an individual activity. Availability of communication networks like Internet will stimulate this collaboration.
- *Adaptive technology.* Transparent shifting between different models, but also between different methods will increase the power of computer models. At present, an important bottleneck is to how to decide when a specific technique should be used. Research is needed to formalize selection of application techniques.
- *Switching perspective.* Distinguishing between different views of the model contents is necessary. Future developments of conceptual languages and automated modeling will allow a transparent switching between physical, mathematical and economical viewpoints, etc., and allow propagation of information across different views.

9.4 Tools and techniques

So far we have dealt with the structuring and formalization of models and modeling, and we have showed some recent and expected developments. All these developments depend on what the computer support systems allow us to do. What can we expect from computer science that will allow progress in automated modeling?

As concerns hardware:

- *Growing resources.* Mentioning the expected increase in computational speed and storage capacity is almost needless. Numerical barriers will become a less predominant factor in the modeling process. Note however that increasing computer power adds to the need for proper assumption management, although the opposite is often observed (Truesdell 1984).
- *Networks.* Models will be developed in a distributed way. This refers to both computational resources and to collaboration between modelers and designers.
- *Multimedia.* Automated modeling will be less dependent on the textual form. Graphical representations will become more important. In particular a more realistic view will be used, exploiting the power of multimedia systems. Developments in the field of *virtual reality* are important. This technology, however, implies as well a risk with respect to modeling as it may obscure the problem that is to be solved.
- *Sensor technology.* If models are used to solve 'real world' problems, calibration and validation by means of measurement cannot be avoided. Limitation of what can be measured will be pushed further away. Construction of intelligent sensors by the development of new IC-technology, will merge observations with interpretation.

If we consider software technology, the following trends will affect model building:

- *Object orientation.* Object oriented software engineering is particularly useful for building modeling and simulation software (Rumbaugh et al 1991). Modular systems will allow rapid prototyping and continuous, iterative design processes. Recently so-called *patterns* (Gamma et al 1994) have become popular reusable frames for software design.
- *Database management systems.* Storing and retrieving large amounts of model fragments in model libraries will become possible by development of efficient and accessible repositories. Searching models is then possible using meta-information like explaining applicability of model fragments within a given context, rather than using internal model structure. The modeler can find models through associative searching.
- *Communication.* Model support systems and databases must be made open, so that different systems can share information and knowledge. This implies exchange protocols and interoperability.
- *Graphical User Interfaces.* The interactive character of the modeling task requires immediate feedback from the computer support system, stressing the importance of

graphical modeling languages. Attempts to make incorrect structures can be reported immediately to the modeler. Moreover, simulation results can be reported in a way that is close to our common-sense understanding of the world.

- *Machine learning.* Besides modeling based on explicit conceptual representations, also bottom-up techniques will be further developed. Machine learning techniques (genetic algorithms, neural networks, case based reasoning, etc.) generate black- and grey-box models from data sets. Integration of data-driven and concept-driven techniques will significantly improve model support systems.
- *Knowledge-based systems.* We may expect important impact from developments in knowledge engineering. In particular formalization of domain knowledge is given much attention (Akkermans et al 1993; Breuker and Van de Velde 1994).

9.5 Conclusions

We have attempted to elucidate some developments to be expected, or at least hoped for, with respect to automated modeling. Of course, such an overview can never be complete, if only because we have to rely on what we presently see as an adequate model of the modeling task.

In our view, the following aspects will significantly affect the modeling process in the future:

- Formalization through logic and mathematics.
- Formalization in terms of conceptual languages.
- Structured problem specification.
- Assumption management.

Two problems can be distinguished at present that may interfere. First, parameterization (identification) of models remains a difficult issue. This is due to a mismatch between what *can* be measured and what *should* be measured according to the model. In this respect, both model and measuring system are critical. Second, defining formal models - at the logical-mathematical or at the conceptual level - will be difficult for ill-structured domains. Developing proper languages will be a major challenge in the coming years.

Both problems result from the essential problem in modeling: finding an appropriate trade-off between generality (theory) and specificity (application). Theories provide a basis for quality models, but they are never applicable without limitations. Practical problems require practical solutions. We will need new ways to push the limits of generalization *and* goal-directness. In that way automated modeling will become more accessible and more widely spread, but also stimulate creativity and innovation.

Exercises

1. Consider the pea model from chapter 1. Discuss how *number crunching, hybrid systems, formal verification* and *non-linear analysis* could be useful for the

purposes outlined in that chapter.

2. There is an obvious advantage of using conceptual languages for model development, making computer codes much more transparent. Discuss how for the different global radiation models of chapter 3 these would apply.
3. The hardware and software technologies defined by the author in chapter 9.4 could have interesting consequences. In particular for collecting data and communication. Discuss how these developments could be used to the advantage in developing and applying crop models.
4. Production ecology focuses on competitive production in an environmental friendly way. Modern technological developments could assist in this respect. Discuss the pros and cons of their integration into production ecology.
5. A short reference was made to qualitative simulation. There the real numbers are expressed in terms of positive, negative or zero only. At first sight this seems an appealing simplification, but it appears to be not very useful in practice. Give an intuitive explanation why this will lead to simulation of dynamic systems.
6. An important modelling problem occurs when both continuous *and* discrete processes occur simultaneously. Give an example of such a situation and describe which part of the system is continuous and which should be viewed as discrete (in time).
7. We have seen trends and expectations in computer based modeling. Give your own (speculative) view of how an ideal future modeling tool could support the modeling process, using Fig. 9.1.

References

Addiscott TM 1993 Simulation modelling and soil behaviour. Geoderma 60:15-40

Addiscott TM, Smith JU, Bradbury NJ 1996 Critical evaluation of models and their parameters. J Environ Qual 24:803-807

Addiscott TM, Tuck G 1996 Sensitivity analysis for regional-scale solute transport modelling. In: Corwin DL, Loague K (eds) Application of GIS modelling of non-point source pollutants in the vadose zone. Soil Science Society of America Special Publication No 48. Madison WI: SSSA, pp 153-162

Addiscott TM, Wagenet RJ 1985 A simple method for combining soil properties that show variability. Soil Sci Soc Am J 49:1365-1369

Addiscott TM, Whitmore AP 1987 Computer simulation of changes in soil mineral nitrogen and crop nitrogen during autumn, winter and spring. J Agr Sci 109: 141-157

Addiscott TM, Whitmore AP 1991 Simulation of solute leaching in soils of differing permeabilities. Soil Use Manage 7:14-21

Akkermans JM, Van Harmelen F, Schreiber AT, Wielinga BJ 1993 A Formalisation of knowledge-level models for knowledge acquisition. Int J Intell Syst 8:169-208

Aldenberg T, Janse JH, Kramer PRG 1995 Fitting the dynamical model PCLake to a multi-lake survey through Bayesian statistics. Ecol Model 78:83-99

Ångström A 1924 Solar and terrestrial radiation. Q J Roy Meteor Soc 50:121-125

Anselin L, Getis A 1992 Spatial statistical analysis and geographic information systems. Ann Reg Sci 1:3–23

Anthony S, Quinn P, Lord E 1996 Catchment scale modelling of nitrate leaching. Aspects of Applied Biology 46 (*in prep*)

Attonaty JM, Pasquier G 1996 Multi-agent models: an approach to be explored to improve individual decision-making aids. In: Beers G, Huirne RBM, Pruis HC (eds) Farmers in small-scale and large-scale farming in a new perspective: objectives, decision making and information requirements. Den Haag: LEI-DLO, pp 183-197

Barnsley M 1999 Digital remotely-sensed data and their characteristics. In: Longley PA, Goodchild MF, Maguire DJ, Rhind DW (eds) Geographical information systems: principles, techniques, management and applications. New York: Wiley, 1, pp 449–64

Batey P, Brown P 1995 From human ecology to customer targeting: the evolution of geodemographics. In Longley P, Clarke G (eds) GIS for business and service planning. Cambridge: GeoInformation International: pp 77–103

Batty M 1976 Urban modelling: algorithms, calibrations, predictions. Cambridge, Cambridge University Press

Batty M 1981 Urban models. In: Wrigley N, Bennett R J (eds) Quantitative geography: a British view. London: Routledge and Kegan Paul: pp 181–91

Batty M, Longley P 1994 Fractal cities: a geometry of form and function. London, Academic Press

Bayes T 1764 An essay towards solving a problem in the doctrine of chances. Philos T Roy Soc Lon for 1763, 53:370-418

Berger JO 1985 Statistical decision theory and Bayesian analysis (second edition). New York: Springer

Bernoulli D 1760 Essay d'une nouvelle analyse de la mortalité causée par la petite vérole et les avantages de l'inoculation pour la prévenir. Hist Ac Roy Sci 1-45 (published in 1766)

Berry BJL 1971 Introduction: the logic and limitations of comparative factorial ecology. Econ Geogr 47:209–219

Berry BJL, Marble D F (eds) 1968 Spatial analysis: a reader in statistical geography. Englewood Cliffs NJ: Prentice-Hall

Besselat B, Cour P 1990 La prévision de la production viticole à l'Aide de la technique 'Capture de pollen'. In: Proceedings of the conference on the application of remote sensing to agricultural statistics, Varese, Italy, 10-11 October 1989, pp 261-268. , Luxembourg: Publ EUR 12581 of the Office for Official Publications of the EC

Besselat B, Cour P 1996 Guide pratique: Elaboration d'une prévision de récolte à partir du dosage pollinque de l'atmosphère. Ispra: Publ EUR 16422 de l'Office des Publication Officielles de la Commission Européenne

Birkin M 1996 Retail location modelling in GIS. In: Longley PA, Batty M (eds) Spatial analysis: modelling in a GIS environment. Cambridge: GeoInformation International: pp 207–25

Booch G 1994 Object-oriented analysis and design with applications. Second edition. Redwood City: Benjamin Cummings

Boons-Prins ER, De Koning GHJ, Van Diepen CA, Penning de Vries FWT 1993 Crop specific simulation parameters for yield forecasting accross the European Community. In: Rabbinge R, Ward SA, Van Laar HH (eds) Simulation and Systems Management in Crop Protection. Wageningen: Pudoc

Boote KJ, Jones JW, Pickering NB 1996 Potential uses and limitations of crop models. Agron J 88:704-716

Bories L 1996 The OLIWIN Project: Agro-meteorological models for the estimation of olive and grapevine yield. Internal report. Ispra: Joint Research Centre, Space Applications Institute

Bouman BAM, Schapendonk AHCM, Stol W, Van Kraailingen DWG 1996 Description of the growth model LINGRA as implemented in CGMS. Quantitative Approaches in Systems Analysis 7. Wageningen: CT de Wit Research School for Production Ecology

Box GEP, Tiao GC 1973 Bayesian inference in statistical analysis. Reading, Mass: Addison-Wesley

Bradbury NJ, Whitmore AP, Hart PBS, Jenkinson DS 1993 Modelling the fate of nitrogen in crop and soil in the years following application of ^{15}N-labelled fertilizer to winter wheat. J Agr Sci 121:363-379

Breuker J, Van de Velde W (eds) 1994 CommonKADS library for Expertise Modelling. Amsterdam: IOS-Press

Breunese APJ, Top JL, Broenink JF, Akkermans JM 1998 Libraries of reusable models: Theory and application. Simulation 71:7-22.

Brochet P, Gerbier N, Bedel J 1975 Contribution à l'étude agrométéorologique du mais: applications à la prévision des phases phénologiques et des rendements. Paris: Météorologie Nationale

Brodie ML 1984 On the Development of Data Models. In: Brodie ML Mylopoulus J Schmidt JW (eds) On conceptual modelling. New York: Springer, pp 19-48

Brodie ML, Ridjanovic D 1984 On the design and specification of database transactions. In: Brodie ML, Mylopoulos J, Schmidt JW (eds) On conceptual modelling. New York: Springer, pp 277-306
Burrough PA 1996 Environmental modelling with geographic information systems. In: Stein A, Penning de Vries FWT, Schotman PJ (eds) Models in Action. Quantitative Approaches in Systems Analysis 6:59-69
Burrough PA 1998 Dynamic modelling and geocomputation. In Longley P A, Brooks S M, McDonnell R, Macmillan WD (eds) Geocomputation: a primer. Chichester: Wiley pp 165–91
CEC 1985 Soil Map of the European Communities. Luxembourg: Office for Official Publications and Directorate General VI-Agriculture of the European Communities
Chatfield C 1983 Statistics for Technology third edition. London: Chapman & Hall
Choisnel E, De Villele O, Lacroze F 1992 Une approche uniformisée du calcul de l'évapotranspiration potentielle pour l'ensemble des pays de la Communauté Européenne. Luxembourg: Publ EUR 14223 of the Office for Official Publications of the EC
Clark BD, Gleave MB (eds) Social patterns in cities. Institute of British Geographers Special Publication 5. London: IBG
Cliff A, Ord JK 1973 Spatial autocorrelation. London: Pion
Coad P, Yourdon E 1991a Object-oriented analysis - second edition. Englewood Cliffs NJ: Yourdon Press
Couclelis H 1998 Geocomputation in context. In: Longley PA, Brooks SM, McDonnell R, Macmillan WD (eds) Geocomputation: a primer. Chichester: Wiley pp 17–30
Cox DR, Hinkley DV 1974 Theoretical statistics. London: Chapman & Hall
Cressie N 1991 Statistics for spatial data. New York: John Wiley
Cropper SL, Smith JU, Scholefield D, Brown CD, Hollis JM 1996 An empirical approach to predicting nitrate leached to surface water from catchments of mixed land use. J Agr Sci, in prep
Curran PJ, Milton EJ, Atkinson PM, Foody GM 1998 Remote sensing: from data to understanding. In. Longley PA, Brooks SM, McDonnell R, Macmillan WD (eds) Geocomputation: a primer. Chichester: Wiley pp 33–59
Dahl OJ, Myhrhaug B, Nygaard K 1982 Common base language [SIMULA 67]- fourth edition. Oslo: Norwegian Computing Center
Dayan E, Van Keulen H, Jones JW, Zipori I, Shmuel D, Challa H 1993 Development, calibration and validation of a greenhouse tomato growth model: I. Description of the model. Agr Syst 43:145-163
Dayan E, Van Keulen H, Jones JW, Zipori I, Shmuel D, Challa H 1993 Development, calibration and validation of a greenhouse tomato growth model: II. Field calibration and validation. Agr Syst 43:165-183
De Koning ANM 1994 Development and dry matter distribution in tomato: a quantitative approach. PhD thesis, Wageningen Agricultural University
De Wit CT, van Keulen H 1987 Modelling production of field crops and its requirements. Geoderma 40:253-265
Doorenbos J, Kassam AH 1979 Yield response to water. FAO Irrigation and Drainage Paper N°33. Rome: FAO

Doorenbos J, Pruit WO 1984 Guidelines for predicting crop water requirements. Irrigation and Drainage Paper N° 24. Rome: FAO
Doré T 1992 Etude, par voie d'enquête, de la variabilité des rendements et des effets précédent du pois protéagineux de printemps (*Pisum* sativum L.).Thèse de Doctorat, INA Paris-Grignon
Egenhofer MJ, Frank AU 1989 Object-Oriented Modelling. In GIS: Inheritance and propagation. Auto-Carto 9, p588-598
Elshaw Thrall S, Thrall G 1999 Desktop GIS software. In: Longley PA, Goodchild MF, Maguire DJ, Rhind DW (eds) Geographical information systems: principles, techniques, management and applications. New York: Wiley, 1 pp 329–43
FAO 1986 Early agrometeorological crop yield assessment. FAO Plant Production and Protection Paper N°73. Rome: FAO
Flinn AM, Pate JS 1970 A quantitative study of carbon transfer from pod and substending leaf to the ripening seeds of the field pea (*Pisum arvense* L.). J Exp Bot 21:71-82
Frank AU, Kuhn, W 1986 Cell Graphs: A Provable Correct Method for the Storage of Geometry. Proceedings of the 2nd International Symposium on Spatial Data Handling, Seattle. Williamsville NY: International Geographical Union, pp 411-436
Gamma E, Helm R, Johnson R, Vlissides J 1994 Design patterns: elements of reusable object-oriented software. Reading Mass: Addison Wesley
Gelman A, Carlin JB, Stern HS, Rubin DB 1995 Bayesian data analysis. London: Chapman & Hall
Genovese G 1997 The evaluation of the MARS crop yield forecasting model. Report submitted to the committee for agricultural statistics of the EC. Session of October 1997. Luxembourg: European Statistical Office Eurostat (In print)
Gersting JL 1992 Mathematical structures for computer science - third edition. New York: Computer Science Press
Getis A, Ord K 1992 The analysis of spatial association by the use of distance statistics. Geogr Anal 24:189–206
Goodchild MF 1992 Geographical information science. Int J Geogr Inform Syst 6: 31–45
Goodchild MF 1997 Towards a geography of geographic information in a digital world. Comput Environ Urban 21: 377–91
Goodchild MF 1998 Different data sources and diverse data structures: metadata and other solutions. In: Longley PA, Brooks SM, McDonnell R, Macmillan WD (eds) Geocomputation: a primer. Chichester, Wiley pp 61–73
Goodchild MF, Longley PA 1999 The future of GIS and spatial analysis. In: Longley PA, Goodchild MF, Maguire DJ, Rhind DW (eds) Geographical information systems: principles, techniques, management and applications. New York: Wiley, 1 pp 567–80
Goodchild MF, Parks BO, Steyaert LT (eds) 1993 Environmental modeling with GIS. New York: Oxford University Press
Goodchild MF, Steyaert LT, Parks BO, Johnston C, Maidment D, Crane M, Glendinning S (eds) 1996 GIS and environmental modeling: progress and research issues. Fort Collins: GIS World Inc
Graham I 1994 Object oriented methods - second edition. Reading Mass: Addison Wesley
Gruber TR, Olson GR 1994 An ontology for engineering mathematics. In: Doyle J et al

(eds) Fourth Int. Conf. On Principles of Knowledge Representation and Reasoning. San Mateo CA: Morgan Kaufmann

Gupta SC, Larson WE 1979 Estimating soil water retention characteristics from particle size distribution, organic matter percent, and bulk density. Water Resour Res 15:1633-1635

Haggett P 1965 Locational analysis in human geography. London: Arnold

Hargreaves GL, Hargreaves GH, Riley P 1985 Irrigation water requirement for the Senegal River Basin. J Irrig Drain Eng 111:265-275

Harris R 1999 A comparative analysis of a lifestyle and geodemographic typology. Working paper. Bristol, University of Bristol

Hashimoto Y 1989 Recent strategies of optimal growth regulation by the speaking plant concept. Acta Hortic 260: 115-121

Hassanizadeh SM, Carrera J 1992 Editorial: introduction to special issue on validation. Adv Water Resour 15: 75-83

Haun JR 1982 Early prediction of corn yields from daily weather data and single predetermined seasonal constants. Agric Meteorol 27:191-207

Hesse W, Leahy FJ 1992 Authorative topographic-cartographic information system ATKIS. Bonn: Landes Vermessungamt Nordrhein-Wetsfalen

Hogeweg P, Hesper B 1990 Individual modeling in ecology. Math Comput Model 13:83-90

Howard PJA, Loveland PJ, Bradley RI, Dry FT, Howard DM, Howard DC 1995 The carbon content of soil and its geographical distribution in Great Britain. Soil Use Manage 11:9-15

Iwasa Y, Roughgarden J 1984 Shoot/root balance of plants: optimal growth of a system with many vegetative organs. Theor Popul Biol 25: 78-105

Jacobson I, Christerson M, Jonsson P, Overgaard G 1992 Object-oriented software engineering - A use case driven approach. New York: ACM Press

Jacucci G, Kabat P, Pereira LS and 13 others 1993 HYDRA: A decision-support model for irrigation water management. Proceedings of the 2nd Workshop on Crop-Water Models, ICID 15th Congress on Irrigation and Drainage, The Hague, Netherlands, 17pp

Jansen MJW, Rossing WAH, Daamen RA 1994 Monte-Carlo estimation of uncertainty contributions from several independent multivariate sources. In: Grasman J, Van Straten G (eds) Predictability and non-linear modelling in natural sciences and economics. Dordrecht: Kluwer, pp 334-343

Jeuffroy MH 1991 Rôle de la vitesse de croissance, de la répartition des assimilats et de la nutrition azotée dans l'élaboration du nombre de graines du pois protéagineux de printemps (*Pisum sativum* L.). Thèse de Doctorat, Univ. Paris XI

Jeuffroy MH 1994 Le nombre de graines par tige. Agrophysiologie du pois protéagineux ; applications à la production agricole. INRA, ITCF, UNIP, TEC & DOC Lavoisier, Paris pp 93-110

Jeuffroy MH, Chabanet C 1994 A model to predict seed number per pod from early pod growth rate in pea (*Pisum sativum* L.). J Exp Bot 45:709-715

Jeuffroy MH, Devienne F 1995 A simulation model for assimilate partitioning between pods in pea (*Pisum sativum* L.) during the period of seed set ; validation in field conditions. Field Crops Res 41:79-89

Jeuffroy MH, Duthion C, Meynard JM, Pigeaire A 1990 Effect of a short period of high day temperatures during flowering on the seed number per pod of pea (*Pisum sativum* L.). Agronomie 2: 139-145

Jeuffroy MH, Sebillotte M 1997 The end of flowering in pea : influence of plant nitrogen nutrition. Eur J Agron, 6:15-24

Jeuffroy MH, Warembourg FR 1991 Carbon transfer and partitioning between vegetative and reproductive organs in *Pisum sativum* L. Plant Physiol 97: 440-448

Kabat P, Marschall B, Van Den Broek BJ, Vos J, Van Keulen H 1995 Modelling and parameterization of the soil-plant-atmosphere system: a comparison of potato growth models. Wageningen: Wageningen Pers

Karnopp DC, Margolis DL, Rosenberg RC 1990 System dynamics: a unified approach. NewYork: Wiley

Karr EJ, Linck AJ, Swanson CA 1959 The effect of short periods of high temperature during day and night periods on pea yields. Am J Bot 46: 91-93

Keene S 1989 Object oriented programming in Common Lisp, a programmer's guide to CLOS. Reading Mass: Addison Wesley

Keesman K, Van Straten G 1990 Set membership approach to identification and prediction of lake eutrophication. Water Resour Res 26:2643-2652

Kemp Z 1990 An object-oriented model for spatial data. In: Brassel K, Kishimoto H (eds). Proceedings of the 4th International Symposium om Spatial Data Handling. Zurich: University of Zurich, pp 659-668

King D, Daroussin D, Jamagne M 1994 Development of a soil geographic database from the soil map of the European Communities. Catena 21:37-56

King D, Le Bas C 1994 Programme d'estimation des réserves en eau des sols à partir des paramètres de base de données géographiques des sols de l'Union Européenne. Final report of the Contract N° 5538-93-11 EP ISP F for the EC Joint Research Centre, Ispra

Kompala DS, Ramkrishna D, Jansen NB, Tsao GT 1986 Investigation of bacterial growth on mixed substrates: experimental evaluation of cybernetic models. Biotechnol Bioeng 28: 1044-1055

Kooman PL 1995 Yielding ability of potato crops as influenced by temperature and daylenght. PhD thesis, Wageningen Agricultural University

Krasner GE, Pope ST 1988 A cookbook for using the model-view controller user interface paradigm in Smalltalk-80. J Object-Orient Prog 1:26-49

Kroese AH, Van Der Meulen EA, Poortema K, Schaafsma W 1995 Distributional inference. Stat Neerlandica 49, 63-82

Lambert RG, Linck AJ 1958 Effects of high temperature on yield of peas. Plant Physiol 33: 347-350

Lange A and Gilbert C 1999 Using GPS for GIS data capture. In: Longley PA, Goodchild MF, Maguire DJ, Rhind DW (eds) Geographical information systems: principles, techniques, management and applications. New York: Wiley, 1 pp 465–74

Leenhardt D 1995 Errors in the estimation of soil water properties and their propagation through a hydrological model. Soil Use Manage 11:15-21

Leffelaar PA 1988 Dynamics of partial anaerobiosis, denitrification, and water in a soil aggregate: simulation. Soil Sci 146:427-444

Leffelaar PA 1990 On scale problems in modelling: and example from soil ecology. In:

Rabbinge R et al (eds) Theoretical production ecology: reflections and prospects. Wageningen: Pudoc
Levine G 1981 Introductory statistics for psychology: the logic and the methods. New York: Academic Press
Lewis FL 1986 Optimal Control. New York: Wiley
Ljung L, Glad T 1994 Modeling of Dynamic Systems. Englewood Cliffs NJ: Prentice Hall
Loague K, Green RE 1991 Statistical and graphical methods for evaluating solute transport models: Overview and application. J Contamin Hydrol 7:51-73
Longley PA 1998 GIS and the development of digital urban infrastructure. Environ Plan B, Anniversary Issue: 53–6
Longley PA, Batty M 1996 Analysis, modelling, forecasting and GIS technology. In: Longley PA, Batty M (eds) Spatial analysis: modelling in a GIS environment. Cambridge: GeoInformation International pp 1–15
Longley PA, Brooks SM, McDonnell R, Macmillan WD (eds) 1998 Geocomputation: a primer. Chichester: Wiley
Longley PA, Goodchild MF, Maguire DJ, Rhind DW 1999 Introduction. In: Longley PA, Goodchild MF, Maguire DJ, Rhind DW (eds) Geographical information systems: principles, techniques, management and applications. New York: Wiley pp 1–20
Longley PA, Mesev V 1997 Beyond analogue models: space filling and density measures of an urban settlement. Papers in Regional Science 76: 409–27
Loudjani P 1995 Remote Sensing and Crop Yield estimation: an Overview. COST 77 Document. Luxembour: Publ EUR 16275 of the Office for Official Publications of the EC
Mandelbrot BB 1986 Multifractals and fractals. Physics Today: September 11–12
Marascuilo LA, Levin JR 1983 Multivariate statistics in the social sciences: a researcher's guide. Monterey CA: Brooks/Cole
Martinez Casanovas JA 1994 Hydrographic information abstraction for erosion modelling at regional level. MSc thesis, Dept of Landsurveying and Remote Sensing, Wageningen Agricultural University, Wageningen
Marx RW 1990 The TIGER system: automating the geographic structure of the United States census. In: Peuquet DJ, Marble DF (eds) Introductory readings in GIS. London: Taylor and Francis, pp 120-141
Matheron G 1971 The theory of regionalized variables and its applications. Paris: Ecole National Superieure des Mines
Meteo Consult 1991 AMDaC User Manual. Software package for Actual Meteorological Database Construction. Wageningen: Meteo Consult BV
Metselaar K 1999 Auditing predictive models: a case study in crop growth. PhD thesis, Wageningen Agricultural University
Meyer B 1988 Object-Oriented Software Construction. Englewood Cliffs NJ: Prentice Hall
Meyer B 1995 On the role of methodology. In: Alagar VS, Missaoni R (eds) Object-oriented technology for databases and software systems. World Scientific, Singapore, pp 1-6
Miles AR, Turner STD, Lord EI 1995 National agricultural pollution modelling in a GIS environment. 7th Annual UK ESRI User Conference, Nottingham University

Molenaar M 1989 Single valued vector maps - a concept in GIS. Geo-Informations-Systeme 2:18-26
Molenaar M 1993 Object hierarchies and uncertainty in GIS or Why is standardization so difficult. Geo-Informations-Systeme 6: 22-28
Molenaar M 1994 A syntax for the representation of fuzzy spatial objects. In: Molenaar M, De Hoop S (eds) Advanced geographic data modelling. Netherlands Geodetic Commission, New Series, Nr40, Delft, pp155-169
Molenaar M. 1998 An introduction to the theory of spatial object modelling for GIS. London: Taylor and Francis
Monarchi D, Puhr GI 1992 A research typology for object-oriented analysis and design. Commun ACM 35(9):35
Monteith JL 1996 The quest for balance in crop modeling. Agron J 88: 695-697
Motha RP, Heddinghaus TR 1986 The joint agricultural weather facility's operational assessment program. B Am Meteorol Soc: 67:1114-1122
Moxey A, McClean C, Allanson P 1995 Transforming the spatial basis of agricultural census cover data. Soil Use Manage 11: 21-25
Munack A 1989 Optimal feeding strategy for identification of Monod-type models by fed-batch experiments. In: Fish NM, Fox RI, Thornhill NF (eds) Computer applications in fermentation technology: modelling and control of biotechnological processes. Amsterdam: Elsevier, pp 195-204
Munier-Jolain NG 1994 Etude de la variabilité du poids individuel des graines du soja de type indéterminé (*Glycine max.* L. Merrill, cv Maple Arrow). Influence de l'apparition séquentielle des organes reproducteurs. Thèse de Doctorat, INRA Paris-Grignon
Ney B, Duthion C, Fontaine E 1993 Timing of reproductive abortions in relation to cell division, water content and growth of pea seeds. Crop Sci 33: 267-270
Ney B, Turc O 1993 Heat-unit-based description of the reproductive development of pea. Crop Sci 33: 510-514
Openshaw S 1998 Building automated geographical analysis and explanation machines. In: Longley PA, Brooks SM, McDonnell R, Macmillan WD (eds) Geocomputation: a primer. Chichester: Wiley pp 95–115
Openshaw S, Openshaw C 1997 Artificial intelligence in geography. Chichester: Wiley
Palm R, Dagnelie P 1993 Tendance et effets du climat dans la prévision des rendements agricoles des différents pays de la CE. Luxembourg: Publ EUR 15106 of the Office for Official Publications of the EC
Pancake CM 1995 The promise and cost of object technology: a five year forecast. Commun ACM 38(10):33-49
Paynter HM 1961 Analysis and design of engineering systems. Cambridge Mass: MIT-press
Pigeaire A, Duthion C, Turc O 1986 Characterization of the final stage in seed abortion in indeterminate soybean, white lupin and pea. Agronomie 6: 371-378
Place RE, Brown DM 1987 Modelling corn yields from soil moisture estimates: description, sensitivity analysis and validation. Agric For Meteorol 41:31-56
Powlson DS, Mirza NA 1996 A decision support system to inform policy making for nitrate management. In: Multiple land use and catchment management. Aberdeen University: Aberdeen Research Consortium, pp 24.

Puckett WE, Dane JH, Hajek BF 1985 Physical and mineralogical data to determine soil hydraulic properties. Soil Sci Soc Am J 49:831-836

Rao PSC, Rao PV, Davidson JM 1977 Estimation of the spatial variability of the soil-water flux. Soil Sci Soc Am Proc 41:1208-1209

Rawls WJ, Brakensiek DL, Saxtom DE 1982 Estimation of soil water properties. T ASAE 25:1316-1320

Reilly PM 1976 The numerical computation of posterior distributions in Bayesian statistical inference. Appl Statist 25: 201-209

Rhind D (ed.) 1997 Framework for the world. Cambridge: GeoInformation International

Rhind D 1999 National and international geospatial data. In: Longley PA, Goodchild MF, Maguire DJ, Rhind DW (eds) Geographical information systems: principles, techniques, management and applications. New York: Wiley, 2 pp 769–89

Rich E, Knight K 1991 Artificial intelligence - second edition. New York: McGraw Hill

Richardson DE 1993 Automatic spatial and thematic generalization using a context transformation model. Ottawa: R&B Publications

Richardson DE 1994 Contextual transformations and generalizations of remotely sensed imagery for map generation. In: Molenaar M, De Hoop S (eds) Advanced geographic data modelling, Netherlands Geodetic Commission, New Series, Nr 40, Delft, pp170-178

Roche R, Jeuffroy MH, Ney B 1998 A model to simulate the final number of reproductive nodes in pea (*Pisum sativum L.*). Ann Bot 81: 545-555.

Rodda HJE, Scholefield D, Webb BW, Walling DE 1995 Management model for predicting nitrate leaching from grassland catchments in the United Kingdom: 1. Model development. Hydrol Sci 40: 433-451

Rosenbluth A, Wiener N 1945 The role of models in science. Phil Sci 12:316-321

Rothenberg J 1989 The nature of modelling. In: Widman LE et al (eds), Artificial intelligence, simulation and modelling. New York: Wiley

Rumbaugh J, Blaha M, Premerlani W, Eddy F, Lorensen W 1991 Object-oriented modelling and design. Englewood Cliffs NJ: Prentice-Hall

Sakamoto CM 1978 The Z-index as a variable for crop yield estimation. Agr Forest Meteorol 19:305-313

Salgé F 1999 National and international data standards. In: Longley PA, Goodchild MF, Maguire DJ, Rhind DW (eds) Geographical information systems: principles, techniques, management and applications. New York: Wiley, 2 pp 693–706

Salter PJ, Berry G, Williams JB 1966 The influence of texture on the moisture characteristics of soils. III Quantitative relationships between particle size, composition and available water capacity. J Soil Sci 17:93-98

Sayer RA 1979 Understanding urban models versus understanding cities. Environ Plan A 11: 853–62

Schmidt DC 1995 Using design patterns to develop reusable object-oriented communication software. Commun ACM 38(10):65-74

Scholefield D, Lockyer DR, Whitehead DC, Tyson KC 1991 A model to predict transformations and losses of nitrogen in UK pastures grazed by beef cattle. Plant Soil 132:165-177

Schweppe FC 1983 Uncertain dynamic systems. Englewood Cliffs NJ: Prentice-Hall

Seginer I, Ioslovich I 1996 Crop model reduction and simulation in reduced space. Acta Hortic 406:63-71

Semenov MA, Barrow EM 1997 Use of a stochastic weather generator in the development of climate change scenarios. Climatic Change 35:397-414

Shiffer M 1999 Managing public discourse: towards the augmentation of GIS with multimedia. In Longley PA, Goodchild MF, Maguire DJ, Rhind DW (eds) Geographical information systems: principles, techniques, management and applications. New York: Wiley, 2 pp 725–33

Shlear S, Mellor SJ 1992 Object-oriented systems analysis: Modeling the world in states. Englewood Cliffs NJ: Yourdon Press

Siegel S 1956 Nonparametric statistics for the behavioral sciences. New York: McGraw-Hill

Smith J, Scholefield D, Glendining M, Stockdale E 1998 Using models to optimise the efficiency of nitrogen across whole farm rotations. In: Van Keulen H, Lantinga EA, Van Laar HH (eds) Proceedings of an international workshop on mixed farming systems in Europe, Dronten/Wageningen, The Netherlands 25-28 May 1998. Wageningen: Ir AP Minderhoudhoeve-series no. 2. pp 151-157

Smith JU 1995a Nitrate Losses from Arable Land in Chichester Harbour: Initial Estimates using the SUNDIAL Model. In: Chichester Harbour land use and nitrates study. Report for National Rivers Authority (Sussex Area) and Chichester Harbour Conservancy (MJ Banham)

Smith JU 1995b Systematic determination model parameters. Application to field scale simulation of nitrogen dynamics in arable land. Proceedings of the British Society of Soil Science Annual Meeting, September, 11-14, 1995

Smith JU, Bradbury NJ, Addiscott TM 1996a SUNDIAL: A PC-based version of the Rothamsted nitrogen turnover model. Agron J (in press)

Smith JU, Glendining MJ 1996 A decision support system for optimising the use of nitrogen in crop rotations. Rotations and cropping systems, Asp Appl Biol 47:103-110

Smith JU, Glendining MJ, Smith P 1997a The use of computer simulation models to optimse the use of nitrogen in whole farm systems. Optimising cereal inputs: its scientific basis, Asp Appl Biol 50:147-154

Smith JU, Smith P, Addiscott TM, 1996b Quantitative methods to evaluate and compare soil organic matter (SOM) models. In: Powlson DS et al (eds) Evaluation of soil organic matter models using existing long-term datasets. NATO ASI Series I, 38:181-200

Smith P, Smith JU, Powlson DS and 17 others 1997b A comparison of the performance of nine soil organic matter models using seven long-term experimental datasets. Geoderma 81: 153-225

Spitters CJT 1987 An analysis of variation in yield among potato cultivars in terms of light absorption, light utilisation and dry matter partitioning. Acta Hortic 214:71-84

Spitters CJT 1990 Crop growth models: their usefulness and limitations. Acta Hortic 267: 349-368

Spitters CJT, Van Keulen H, Van Kraalingen DWG 1989 A simple and universal crop growth simulator: SUCROS87. In: Rabbinge R, Ward SA, Van Laar HH (eds) Simulation and Systems Management in Crop Protection. Wageningen: Pudoc, pp 147-181

Stein A, Bouma J, Mulders MA, Weterings MHW 1989 Using co-kriging in variability studies to predict physical land qualities of a level river terrace. Soil Technol 2:385-402

Strömberg JE, Top JL, Söderman U 1993 Variable causality caused by discrete effects. In: Granda JJ, Cellier FE (eds) Proceedings of ICBGM'93. San Diego: The Society for Computer Simulation, pp 115-119

Stroustrup B 1991 The C++ Programming Language (2nd edition). Reading Mass: Addison Wesley

Supit Y 1994 Global radiation. Luxembourg: Publ EUR 14767 of the Office for Official Publications of the EC

Supit Y, Hooijer AA, Van Diepen CA (eds) 1994 System description of the WOFOST 6.0 crop simulation model implemented in CGMS. Luxembourg: Publ EUR 15956 of the Office for Official Publications of the EC

Swanson ER, Nyankori JC 1979 The influence of weather and technology on corn and soybean yield trends. Agric Meteorol 20:327-342

Szynkier K 1974 The effect of removing of supply or sink organs on the distribution of assimilates in two varieties of garden pea, *Pisum sativum* L. Acta Agr Scand 24: 7-12

Tap RF, Van Straten G 1995 Development of a reduced order tomato model. 1st IMACS/IFAC Symposium on Mathematical modelling and simulation in agriculture & bio-industries. Vol III VI.A2-1:6. Brussels: Free University of Brussels

Te Braake HB, Van Straten G 1995 Random activation weight neural net (RAWN) for fast non-iterative training. Eng Appl AI 8:71-80

Tien BT, Van Straten G 1995 Neural-fuzzy systems for non-linear system identification - orthogonal least squares training algorithms and fuzzy rule reduction. Preprints 2nd IFAC/IFIP/EurAgEng Workshop on Artificial Intelligence in Agriculture, Wageningen, The Netherlands, May 29-31, (IFAC-Elsevier), pp 249-254

Timms D 1969 The urban mosaic: towards a theory of residential differentiation. Cambridge: Cambridge University Press

Top JL 1993 Conceptual Modelling of Physical Systems. PhD thesis, University of Twente

Top JL Akkermans JM 1994 Tasks and Ontologies in Engineering Modelling. Int J Human-Computer Studies 41:585-617

Truesdell, C An idiot's fugitive essays on science: methods, criticism, training, circumstances. New York: Springer-Verlag

USBUREAU 1990 Technical description of the DIME System. In: Peuquet DJ, Marble DF (Eds) Introductory readings in GIS, London: Taylor and Francis, pp100-111

USDA 1987 Weekly weather and crop bulletin: National Weather Summary. 74, No40 Washington: US-Departments of Commerce and of Agriculture

Van Bergeijk J, Goense D, Wigboldus K 1995 Object oriented analysis and design of real-time systems: an application of the Shlear-Mellor method for the development of a side computer (in Dutch). Agro-informaticareeks, 9 June 1995

Van der Voet P, Van Diepen C, Oude Voshaar J 1993 Spatial interpolation of daily meteorological data: a knowledge based procedure for the regions of the European Communities. Wageningen: The Winand Staring Centre

Van Diepen CA, Wolf J, Van Keulen H, Rappoldt C 1989 WOFOST: a simulation model of crop production. Soils Use Manage 5:16-24

Van Dijk J, Breedveld PC 1991 Simulation of system models containing zero-order causal paths - I. Classification of zero-order causal paths. J Franklin I 328:959-980
Van Henten EJ, Bontsema J 1995 Non-destructive crop measurements by image processing for crop growth control. J Agric Eng Res 61:97-105
Van Keulen H 1975 Simulation of water use and herbage growth in arid regions. Wageningen: Pudoc
Van Keulen H, Wolf J 1986 Modelling of agricultural production: weather, soils and crops. Wageningen: Pudoc
Velten K, Richter O 1995 Optimal root/shoot-partitioning of carbohydrates in plants. B Math Biol 57:99-107
Vossen P 1990a Comparative statistical validation of two ten day water use models and three yield reduction hypotheses for yield assessment in Botswana. Agr Forest Meteorol 51:177-195
Vossen P 1990b Modèles agrométéorologiques pour le suivi des cultures et la prévision de rendements des grandes régions des communautés Européennes. In: Proceedings of the conference on the application of remote sensing to agricultural statistics, Varese, Italy, pp 75-84. Luxembourg: Publ EUR 12581 of the Office for Official Publications of the EC
Vossen P 1992 Forecasting national crop yields of EC countries: the approach developed by the Agriculture Project. In: Proceedings of the second conference on the Application of remote sensing to agricultural statistics, Belgirate, Italy, pp 159-176. Luxembourg: Publ EUR 14262 of the Office for Official Publications of the EC
Vossen P, Rijks D 1995 Early crop yield assessment of the EC countries: the system implemented by the Joint Research Centre. Luxembourg: Publ EUR 16318 of the Office for Official Publications of the EC
Wallach D, Goffinet B 1989 Mean squared error of prediction as a criterion for evaluating and comparing system models. Ecol Model 44: 299-306
Warrick AW, Mullen GJ, Nielsen DM 1977 Predictions of the soil water flux based upon field-measured soil-water properties. Soil Sci Soc Am J 41:14-19
Warrick AW, Zhang R, Moody MM, Myers DE 1990 Kriging versus alternative interpolators: errors and sensitivity to model inputs. In: K Roth et al (eds) Field-scale solute and water transport through soil. Basel: Birkhäuser
Webster R, Oliver MA 1990 Statistical methods in soil and land resource survey. New York: Oxford University Press
Weld DS, De Kleer J (eds) 1990 Readings in Qualitative Reasoning about Physical Systems. San Mateo CA: Morgan Kaufmann
Whisler FD, Acock B, Baker DN, Fye RE, Hodges HF, Lambert JR, Lemmon HE, Mc Kinion JM, Reddy VR 1986 Crop simulation models in agronomic systems. Adv Agron: 141-208
Whitmore AP 1991 A method for assessing the goodness of computer simulations of soil processes. J Soil Sci 42:289-299
Wilkie G 1993 Object-oriented software engineering - the professional developer's guide. Reading Mass: Addison Wesley
Wirfs-Brock RJ, Johnson RE 1990 Surveying current research in object-oriented design. Commun ACM 33(9):104-124

Young PC, Lees MJ 1996 Simplicity out of complexity in glasshouse climate modelling. Acta Hortic 406: 15-28
Zeigler BP 1976 Theory of Modelling and Simulation. New York: Wiley

INDEX

MIX
Papier aus verantwortungsvollen Quellen
Paper from responsible sources
FSC® C105338

If you have any concerns about our products,
you can contact us on
ProductSafety@springernature.com

In case Publisher is established outside the EU,
the EU authorized representative is:
Springer Nature Customer Service Center GmbH
Europaplatz 3, 69115 Heidelberg, Germany

Printed by Libri Plureos GmbH
in Hamburg, Germany